Dear Bill,

Hope you enjoy! Merry, merry Christmas.

Love,

Kara + Pete

P.S. Mom will kill you if you sail that far away

A VOYAGE FOR MADMEN

Also by Peter Nichols

Sea Change: Alone Across the Atlantic in a Wooden Boat

Voyage to the North Star

N

A VOYAGE FOR MADMEN

PETER NICHOLS

HarperCollins*Publishers*

S

HarperCollins books may be purchased for educational, business, or sales promotional use. For information, please write: Special Markets Department, HarperCollins Publishers Inc., 10 East 53rd Street, New York, NY 10022.

An extension of this copyright page appears on page 297.

FIRST EDITION

Designed by Sarah Gubkin

Library of Congress Cataloging-in-Publication Data
Nichols, Peter
A voyage for madmen/ Peter Nichols.—1st ed.
p. cm.
Includes bibliographical references.
ISBN 0-06-019764-1
1. Sunday Times Golden Globe race (1968–1969) I.Title.
GV832 .N53 2001
7.97. .14—dc21

01 02 03 04 ❖/RRD 10 9 8 7 6 5 4 3

To the memory of my father, Brayton C. Nichols
For his sister, Cynthia Hartshorn
For my cousin, Matt deGarmo

Everything can be found at sea,
according to the spirit of your quest.

—JOSEPH CONRAD

List of Characters

The nine competitors in the Golden Globe race, and their boats, in order of departure:

JOHN RIDGWAY, 29, captain in the British Army. Rowed across the Atlantic with Chay Blyth in a 20-foot open boat in 1966. Departed Inishmore, Ireland, June 1, 1968. Sloop *English Rose IV,* 30-foot-long twin-keeled fiberglass.

CHAY BLYTH, 27, former British Army sergeant. Ridgway's transatlantic rowing partner. Departed Hamble June 8. Sloop *Dytiscus III,* fiberglass, twin-keeled 30-footer, very similar to *English Rose IV.*

ROBIN KNOX-JOHNSTON, 28, British Merchant Marine captain. Departed Falmouth June 14 in the 32-foot-long ketch *Suhaili,* built of teak in India.

BERNARD MOITESSIER, 45, French sailor-author. Sailed with his wife nonstop from Tahiti to Spain, via Cape Horn, in 1965–1966 aboard his 39-foot-long steel ketch *Joshua.* Departed Plymouth, Devon, August 22 aboard *Joshua.*

LOÏCK FOUGERON, 42, French, manager of a motorcycle

company in Casablanca, Morocco. Friend of Moitessier's. Departed Plymouth August 22 in the 30-foot-long, gaff-rigged steel cutter, *Captain Browne.*

BILL KING, 57, farmer, former British Navy submarine commander. His 42-foot-long, junk-rigged, cold-molded wood schooner, *Galway Blazer II,* was designed and built expressly for a nonstop circumnavigation, but not for a race. Departed Plymouth August 24.

NIGEL TETLEY, 45, Royal Navy lieutenant commander. Sailed in his live-aboard home, a 40-foot-long, 22-foot-wide, plywood trimaran ketch, *Victress.* Departed Plymouth September 16.

ALEX CAROZZO, 36, Italian single-hander who had previously sailed alone across the Pacific, in the 66-foot cold-molded wooden ketch *Gancia Americano* built for the Golden Globe race. He "sailed"—that is, he removed to a mooring at Cowes, Isle of Wight, to continue preparations—on the final deadline date set by the race sponsor, the London *Sunday Times*: October 31. He put to sea a week later.

DONALD CROWHURST, 36, English electronic engineer. His 40-foot-long, ketch-rigged, plywood trimaran, *Teignmouth Electron,* was a modified sister ship to Tetley's *Victress.* He too sailed on October 31, within hours of the *Sunday Times* deadline.

INTRODUCTION

TOWARD THE END OF THE 1960S, as Mankind closed in on its goal of voyaging to the moon, nine men set out in small sailboats to race each other around the watery earth, alone and without stopping. It had never been done before. Nobody knew if it could be.

It was dubbed by its eventual sponsor, the London *Sunday Times,* the Golden Globe race. It was the historical progenitor of modern single-handed yacht racing, to which it bears almost no resemblance. Today, high-tech, multimillion-dollar, corporate-sponsored sailing machines race around the world in 100 days or less. Their captains talk by phone and send e-mail to their families and headquarters ashore. They receive weather maps and forecasts by fax. They navigate using the global positioning system (GPS), their locations determined by satellites and accurate to within yards. These positions are simultaneously transmitted to race organizers ashore. Today's racers cannot get lost or file false reports of progress. If they get into trouble, rescue aircraft can often reach their exact locations in a matter of hours. They may race through the most dangerous waters in the world, but their safety net is wide and efficient.*

The Golden Globe racers sailed in the age before satellites provided pinpoint navigation and verification of positions. Like Captain Cook in the eighteenth century, they navigated by sextant, sun, and stars. Their world at sea was far closer to that earlier age than to ours today. When they sailed, heading for the world's stormiest seas in a motley array of new and old boats, they vanished over the horizon into a true unknown. The only information of their whereabouts and what was happening to them came from their own radio transmissions. In time, the radios broke down; several sailors carried no radios at all. One man, sending reports of tremendous progress that made him appear a likely winner, never, in fact, left the Atlantic Ocean, but tried to fake his passage around the world.

Compared with the yachts of today their boats were primitive and unsophisticated—and small: the living space in which the sailors planned to spend the better part of a year was about the size of a Volkswagen bus.

These men sailed for reasons more complex than even they knew. Each decided to make his voyage independent of the others; the race between them was born only of the coincidence of their timing. They were not sportsmen or racing yachtsmen: one didn't even know how to sail when he set off. Their preparations and their boats were as varied as their personalities, and the contrasts were startling. Once at sea, they were exposed to conditions frightening beyond imagination and a loneliness almost unknown in human experience.

Sealed inside their tiny craft, beyond the world's gaze, stripped of any possibility of pretense, the sailors met their truest selves. Who they were—not the sea or the weather—determined the nature of their voyages. They failed and succeeded on the grandest scale. Only one of the nine crossed the

---

*Still, they can disappear and perish at sea, as did Canadian Gerry Roufs in the 1996 Vendée Globe race.

finish line after ten months at sea and passed through fortune's elusive membrane into the sunny world of fame, wealth, and glory. For the others the rewards were a rich mixture of failure, ignominy, sublimity, madness, and death.

The race was the logical inevitability of the first tentative passage made by a man daring to float across a lagoon on a log: in the end, alone and without stopping, he floated so far that he arrived back at the place from where he started, there being no farther earthbound voyage.

Like the first ascent of Everest, it was a feat without any larger purpose than its own end. But like a trip to the moon, it was a voyage that provided Man with another benchmark of the far reach of his yearning endeavor.

The Golden Globe race happened in a different world, as distant, in terms of our experience with the sea, as Joseph Conrad's. The story of that race now has the feel of an older romance of the sea, a tale of unlikely, heroic, desperate, and tragic characters.

At the time of the Golden Globe race, I was a schoolboy in England. I knew nothing of sailboats and sailing. But a few years later I took a brief (disastrous, frightening, and wildly exciting) trip aboard an old wooden schooner and my life derailed and spun away seaward. I spent a decade and a half seriously afloat. I worked my way up from paint-scraping grunt to licensed professional yacht captain, delivering sailboats for owners in the Mediterranean, the Caribbean, and across the Atlantic. Eventually my wife and I bought our own small wooden sailboat and moved aboard it full-time.

During those years I collected and read every book I could find about the sea and small-boat voyaging. The literature of the sea, I found, interested me as much as the sea itself. In time I came across a few books about the Golden Globe race and became fascinated—obsessed—by this story. I scoured newspaper libraries for articles about the race. I wondered what it was like to be alone at sea for long stretches of time, and I wondered about those men. I decided I wanted to try single-handed sailing, to get a little taste of what they had experienced, alone and far out at sea.

I got it—more than I bargained for. After the breakup of my marriage, I started across the Atlantic, bound from England to the United States, alone in my 27-foot-long, 44-year-old wooden sailboat, *Toad*. It was an eventful, bittersweet voyage that ended with *Toad* sinking a week short of reaching the American shore.

Reading of the Golden Globe race as I learned to sail, the story became a core ingredient of my fascination with small boats and the sea. Crossing most of an ocean alone and having my boat founder beneath me only intensified my obsession. This book is the result of a deep investigation into that race, and my efforts to put myself aboard each of those boats and into the minds of those nine very different men.

A VOYAGE FOR MADMEN

I

In 1966–1967, a 65-year-old Englishman, Francis Chichester, sailed alone around the world. He stopped only once, in Australia.

A tall, thin, balding man with thick-lensed glasses, Chichester looked more like a prep school headmaster than an adventurer. He owned a small book and map store in London. He was a vegetarian. But the urge to subject himself to extreme tests characterized his life. In his youth he made a pioneering flight in a small aircraft from England to Australia. In 1960, at the age of 59, he and four friends made a wager to race each other single-handedly in their four very different boats across the Atlantic. The course began at Plymouth's Eddystone Lighthouse and finished at the Ambrose light vessel off New York Harbor; the route between these two points was up to the racers. There were no other rules. The winner would receive half a crown.

Francis Chichester won the bet and the race. Sailing his 39-foot sloop *Gypsy Moth III,* the largest boat of the five, he made Ambrose in 40 days. But winning was not enough; he thought he could do it faster. Two years later, racing nobody but himself, he crossed the Atlantic again, cutting more than 6 days off his earlier voyage. Still he was not happy with his time; he believed a crossing of less than 30 days was possible.

The London *Observer* had covered the 1960 race and found that it owned a story with major and growing public interest. Four years later, in 1964, the *Observer* sponsored a second single-handed transatlantic race (now famously known by its acronym, OSTAR). Ten additional competitors joined the original group. One of the newcomers, the Frenchman Eric Tabarly, trounced the fleet and took the honors in 27 days, 3 hours, 56 minutes. Chichester came second, 20 hours and 1 minute later. He beat his personal target time handily, but second was a new place for him, an ignominious position for a lone adventurer.

Tabarly was awarded the Legion of Honor and became a national hero in France: "Thanks to him it is the French flag that triumphs in the longest and most spectacular race on that ocean which the Anglo-Saxons consider as their special domain," proclaimed the *Paris Jour.*

Single-handed racing hit the big time. National pride on both sides of the English Channel, from two nations famous for their sense of superiority, xenophobia, and rivalry, now focused on the third OSTAR, due to be held in 1968. At least forty sailors announced plans to compete. Many had new, experimental craft designed and built solely for the purpose of winning that one race. Eric Tabarly was building a new 67-foot trimaran, capable of tremendous speeds; at the time this was a radical reappraisal of the size of boat one person could handle. These boats, with their size and gear and engineering, became so expensive that they were beyond the reach of ordinary sailors. Yacht racing began to resemble motor racing, and the long, increasingly ugly hulls were plastered with commercial logos.

A few sailors felt this was veering too far from the notion of "sport." They wrote disapproving letters to yachting magazines, dropped away, and left the field to younger sailors who were learning to navigate the tide rips and currents of commercial sponsorship.

Chichester decided not to compete with the pack in 1968. He would be up against younger men sailing larger boats, and the outcome must have been clear to him: he would be the game old campaigner who would manage a respectable placing halfway through the fleet. He quietly set off to do something else.

Sailing alone around the world was nothing new. The Nova Scotian-born American Joshua Slocum, a sailing ship master beached in his middle years by the steam age, was the first to do it, in 1895–1898. He sailed from Gloucester, Massachusetts, west-about around the globe, against ferocious prevailing winds through the Strait of Magellan, north of Cape Horn, in a seemingly unhandy, fat-hulled, engineless old oystering sloop that he had rebuilt himself and christened *Spray.* The *Spray*'s seagoing abilities, and what Slocum managed to do with her, have been wondered at and argued over by sailors ever since. Slocum (who couldn't swim and nearly drowned trying to set an anchor off the Uruguayan coast) stopped in many places and wrote a drily humorous yet thrilling book of his adventure, *Sailing Alone Around the World.* One hundred years later it is still the standard by which all other sailing narratives are gauged.

Eighteen other men had circumnavigated alone by the time Chichester set out in 1967, but his voyage caught the public imagination as perhaps none other since Slocum's. It was no pleasure cruise. His route was down the Atlantic, east-about around the bottom of the world, back up the Atlantic. Virtually all the east-to-west part of his circumnavigation took place in a sea not found on most atlases but infamously known to all sailors as the Southern Ocean: the windswept southerly wastes of the Atlantic, Pacific, and Indian Oceans between latitudes 40 and 60 degrees south, between the habitable world and the Antarctic, where storm-force westerly winds develop and drive huge seas around the globe, unimpeded by land except at one fearsome place, Cape Horn, the southernmost rock of the Andes, the scorpion-tail tip of South America.

Sailors have respectfully and fearfully labeled the latitudes of this global band of turbulent water the Roaring Forties, the

Furious Fifties, the Screaming Sixties. The tea clippers from India and China and the square-rigged grain ships from Australia took this route back to Europe because, blown by the westerlies through the desolate seas of the Forties and Fifties, circling the planet at a short, high latitude, it was the fastest way around the world.

But it took sailors through the most isolated area of the globe, the emptiest expanse of ocean, the remotest place from land. To take this lonely shortcut, ships and sailors made a Faustian bargain on every trip. They exchanged sea miles for an almost certain hammering by the largest seas on Earth, the stormiest weather. Giant waves, sometimes over 100 feet high; sleet, hail, and snow; icebergs and fog were the conditions that could be expected in any season. Many ships disappeared in the Southern Ocean; many sailors were washed overboard, almost always unrecoverably.

In one place all these terrors were magnified and concentrated into a ship's single greatest trial. At 57 degrees south, Cape Horn forced ships to their farthest, coldest, stormiest south in order to pass into the Atlantic Ocean. Here, Southern Ocean winds and waters are funneled through a relatively narrow gap, Drake Strait, the 600-mile-wide sea passage between Cape Horn and the Antarctic peninsula. The sea bottom shoals off the Horn, raising the already enormous waves, and williwaws of hurricane-force winds scream down off the Andean glaciers; wind, towering waves, and ferocious currents collide, turning Cape Horn waters into a maelstrom. For 400 years, until the construction of the Panama Canal, Cape Horn was a ship's most convenient exit from the Pacific, and many never made it out. It became known as the "graveyard of the sea."

Once around the Horn, square-rig sailors proudly called themselves "Cape Horners"; and those whose fancy permitted it could then wear with pride a single gold ring through the left, portside, ear—the ear that had passed closest to the Horn while bound east out of the Pacific.

This was Chichester's route around the world. His voyage

was a savage intensification of the trials faced by a transatlantic single-hander. His stated purpose was to beat the times of the old sailing ships; alone he would race them in a small modern yacht. It was a simple concept, dangerous and daring, and Chichester significantly upped the ante by stopping only once, in Australia. Not only sailors, but the greater mass of the non-sailing public understood perfectly what was really going on here: it was an ordeal of the first magnitude. It was like climbing Everest alone.

For the British in particular, whose stature on the world stage had been severely reduced since the Churchillian glory of World War II, who had no plucky astronauts, whose government had recently been scandalized by an association between politicians and two prostitutes who had also been sharing their favors with the KGB, Chichester represented a longed-for but not forgotten ideal of heroic endeavor. British newspapers carried front-page photos of the deeply reefed *Gypsy Moth IV* battling gales off Cape Horn (taken by British warships and aircraft standing by, to Chichester's annoyance, to keep an eye on what had suddenly become a national interest). A quarter of a million people filled Plymouth Harbor on the May evening he sailed home. National television schedules were abandoned to cover the event live, and an entire nation watched hour after hour of *Gypsy Moth IV* sailing the last miles through a great fleet of ships and local boats that stretched from shore far out to sea, waiting through the long English twilight to see the lone sailor step ashore. (Television commentators, speculating on his first rocky steps ashore after months at sea, raised the unseemly possibility that the 65-year-old hero might fall flat on his face with the nation looking on. They mused on air about the probability that the welcoming dignitaries would grab him at the earliest moment and prop him up at all costs. In the end, Chichester acquitted himself well, stepping ashore as if getting out of a golf cart.)

Later, in a ceremony consciously echoing the knighting of Sir Francis Drake by Queen Elizabeth I at Plymouth Hoe 400 years earlier, Chichester stepped ashore at Greenwich, London, and knelt before Queen Elizabeth II, who dubbed him with a sword and granted him a knighthood. It was a masterstroke in a jaded era: every Briton knew the scene from school history books; here was myth made real on television, and an intense frisson of national pride swept across the land.

Chichester did not beat the clipper ships' record sailing times, but nobody cared. He was a national hero even before he was knighted by the queen. His book of his voyage, *Gypsy Moth Circles the World,* published that same year, was an instant and lasting best-seller. What he had done had thrilled the public and resurrected glory for the diminished island race.

Why had he done it? Chichester didn't give a hoot about beating the clipper ships—it was just an excuse to go. It was the glib answer he needed when people asked why. The comparison of passage times between a yacht and a clipper ship is the sort of dry, dull detail that might interest a naval architect, historian, maybe even a few sailors, but this wasn't what whipped millions to a national frenzy and drove a middle-aged man to risk his life. Chichester didn't care *why.* He only knew that he had to go.

In his book *The Ulysses Factor* (published just before the Golden Globe race), British author J. R. L. Anderson writes about the lone hero figure in society, the rare character who by his or her exploits stimulates powerful mass excitement. Homer's Ulysses is the classic archetype. Anderson believes this "Ulysses factor"—a powerful drive made up of imagination, self-discipline, selfishness, endurance, fear, courage, and perhaps most of all, social instability—is a genetic instinct in all of us, but dormant in most. Yet we respond strongly and vicariously to the evidence of it in the few whom this instinct drives to unusual endeavors.

"The Devil drives," was Victorian explorer Sir Richard Burton's inadequate answer to why he persisted in going off to Africa and Asia, enduring great hardship, disfiguring pain (a spear through his face), and a constant threat of violent death when he could have stayed home in England and risen to prominence in any number of ways. He felt the urge to be off, to test himself to the brink of tolerance, and he was unable to resist. Brave, indomitable, elusive and unrestrainable, women found him irresistible, men admired him, the public hungrily consumed every account of his exploits. The lone hero of myth and stories from all ages and cultures, described by Joseph Campbell in his book *The Hero With a Thousand Faces,* is a character driven by the Ulysses factor. So is the movie cowboy: a romantic, socially unstable character who appears at the fringes of town, throwing men and women into turmoil before satisfying an underlying social need and then disappearing. His motives are entirely personal; he acts selfishly in his own interest, but his actions have a profound effect on the society around him. Polar explorers Peary, Scott, and Amundsen; Charles Lindbergh, who made the first transatlantic solo flights; mountain climbers; and single-handed sailors—they are all archetypes of the lone hero.

Part of the attraction of these loners is that they invariably look and sound normal: they look like us. They're usually modest when asked how they survived their terrible ordeals, they readily admit their fear, and in so doing they fool the rest of us into thinking that they are like us—or more accurately, that we could be like them. They become our idealized selves, and so they take us with them, in a way, when they climb Mount Everest or sail around Cape Horn.

But they can't answer the question why. They can't make people who couldn't do what they do understand. When asked, before he disappeared on Everest, why he wanted to climb the mountain, George Mallory gave what is still perhaps the best answer, as simple as the solution to a Zen koan: "Because it's there," he said.

The mass adulation provoked by Chichester's voyage, the inspiration he provided the men who sailed in the Golden Globe race, and the fervent efforts of those who readily aided them in risking their lives are clearly responses to this Ulysses factor. Sailors and would-be hero-adventurers everywhere saw what Chichester had reaped in spades—fame and money—and they were aware of what still remained to be accomplished.

In March 1967, while Chichester was still two months from home, Robin Knox-Johnston, a 28-year-old English merchant marine officer, was on leave at his parents' home in Downe, Kent, before joining the merchant ship *Kenya* as its first officer. One morning, his father read of OSTAR victor Eric Tabarly's new trimaran in a newspaper. Over breakfast they speculated about the Frenchman's plans. Knox-Johnston didn't think the boat would be a good choice for the transatlantic. His father suggested Tabarly might be thinking of another circumnavigation.

"I wonder if he is going to try and beat Chichester's time, or perhaps even go round nonstop," said Knox-Johnston senior. "That's about all there's left to do now, isn't it?"

After his father left for work, Robin Knox-Johnston sat at the kitchen table, stirring his coffee and pondering what had just been said. Sooner or later, someone was bound to do exactly that: sail around the world alone, nonstop. Tabarly could pull it off, but the idea of his winning another big sailing prize rankled. "Frenchman Supreme on the Anglo-Saxon Ocean" the French papers had proclaimed when Tabarly had won the OSTAR, and it had infuriated the young Englishman. "By rights," he thought, "a Briton should do it first."

By rights? In England at that moment, young people were marching against the bomb; in the United States, they were protesting the Vietnam War. It was the sixties, a time when young people were turning, virulently and often violently, against the establishment. Not Robin Knox-Johnston. He was the same age as the Beatles and the Rolling Stones, but he was an unfashionable, almost eccentrically square young man from another era. His heroes were Drake and Frobisher, Elizabethan England's famous sea-dog privateers, whose exploits form venerated chunks of the history syllabus absorbed by English schoolchildren, whose land-grabbing, pillaging, and slaughter were characterized by an imperious and peculiarly English sense of God-given superiority. This tradition, this assumption of the English moral right to trailblaze and conquer, characterized Captain Robert Scott's sense of rightful claim to the South Pole. He and all England were affronted by the sudden arrival in the Antarctic of Roald Amundsen and his crack polar crew who, in 1912, beat Scott to the Pole by a month, dogsledding there and back like extreme excursionists, without a man lost. Amundsen, to England's horror and revulsion, had eaten his dogs one by one en route back—he had been pulled to the Pole by his rations; brilliant perhaps, but he was a cold, ruthless, foreign rotter. Scott and his team, burdened by their romantic notions of man-hauling their sleds, all died on the return from the Pole. An epic bungler, Scott was afterward portrayed to generations of schoolchildren in England as the apotheosis of the hero. In England, he won by dying nobly, beaten by the unscrupulous, dog-eating, trespassing Norwegian. *Dieu et mon droit*—"God and my right"—is the motto on the royal arms of England. *By rights.*

At 17 Knox-Johnston flunked the exams for the Royal Navy, so he apprenticed himself to the British India Steam Navigation Company and joined the merchant service instead. He learned knots and splices and marlinspike seamanship unchanged from the time of Nelson. He learned to navigate by sextant, as had Captains Cook and Bligh. He acquired his sailorly arts aboard ships running between London and ports in East Africa, India, and the Persian Gulf— ports that were still outposts of the British Empire, in spirit if no longer in fact. This was the classic POSH route taken by the Empire builders of the British raj, so called because (the favored, more expensive cabins were on the shaded side of the ship: *port* side going *out, starboard* coming *home.*) But probably no place the young seaman sailed to preserved this vanishing world as authentically as aboard the insular, tradition-steeped ships of the British Merchant Marine, which produced sailors and officers as hidebound in their ways as Old Etonians. It was a tough, exacting, nineteenth-century British seaman's training.

While stationed in Bombay, Knox-Johnston and a fellow officer decided to commission a local Indian boatyard to build them a yacht. They sent away to a design office in England for the plans of a sleek ketch they had seen in a yachting magazine, but what came back in the mail was a much older, slower design, a tubby, bluff-bowed double-ender, drawn by American yacht designer William Atkin in 1924 for *Motor Boat* magazine. Based on a type of Norwegian lifeboat known as a *redningskoite,* it was an indisputably seaworthy shape of hull, but a long way from a slick design in a modern yachting magazine. However, time being short, and the plans in hand appearing sturdy and seaworthy, Knox-Johnston and his partner went ahead with the older design.

The Indian carpenters used adzes, axes, and hand-powered bow drills, the same tools and techniques used to build a dhow. The boat was entirely and massively built (overbuilt, modern yacht designers and builders would say) of Indian teak. Her construction and finish were more like a tugboat's than a yacht's. She was christened *Suhaili,* the name given to a southeasterly wind in the Arabian Gulf, and at her launch a coconut was cracked open on her bow while the men who built her chanted ancient blessings.

A third officer had bought a share in the boat a-building, but life interrupted their sailing plans, as always seems to happen around boats. Knox-Johnston's two partners left the project, and his marriage fell apart, perhaps a casualty of the long absences from home that are a fact of the professional seaman's life. Yet through these upheavals he managed to hold onto his new boat. He bought out his ex-partners, and in 1966 he sailed *Suhaili* from India to South Africa in stages with his brother and another merchant officer, all three working at jobs ashore during stops. Then, in November of that year, they set sail for England. Their final passage was a nonstop, 8,000-mile, 74-day voyage from Cape Town to London.

Suhaili had proved herself seaworthy, but as a boat in which to race alone, nonstop, around the world—possibly against Tabarly's 67-foot trimaran—Knox-Johnston thought she was all wrong. Anybody would have agreed with him.

Could he possibly do it? He began imagining what sort of

boat he would need for a voyage of seven to ten months at sea. He wondered too if he could stand being alone all that time without anyone to talk to. A sociable man from a happy middle-class family, with brothers and a sister, the longest he had ever been on his own was 24 hours. Maybe he'd go mad. Such a voyage, he knew, would resemble the most brutal prison sentence: solitary confinement with the hardest of labor and the constant possibility of death by drowning. But he soon realized he didn't care. He wanted to go.

The decision was almost instantaneous. He saw the shape of the voyage, and he wanted to do it. For him, and for the others who would come to the same decision, there was no deliberation, no deeper rationale or reason. The process was identical in each case: once the idea was grasped, the decision was made. Let others reason why.

In early April 1967, while Francis Chichester was still far down in the South Atlantic, seven weeks from home, Robin Knox-Johnston put *Suhaili* up for sale and went to see Colin Mudie, a preeminent English yacht designer, to talk about a boat for a round-the-world voyage. Mudie was enthusiatic and began sketching a boat as they talked.

Chichester's voyage, with its single stop, provoked the same musings in others. A gauntlet had been thrown down to adventurous dreamers everywhere, and a zeitgeist fantasy of a nonstop solo circumnavigation spread through the sailing world. Many people talked about it throughout 1967—at home, in yacht clubs, on weekend cruises, at work—much as Knox-Johnston and his father had done that morning at breakfast. "Someone's bound to do it," they said, and many imagined doing it themselves. For most it remained a pipe dream, but by the time Chichester arrived home in May 1967, at least three other men were making serious plans for nonstop circumnavigations. All of them seemed better prospects for a successful voyage than Knox-Johnston.

The oldest, at 57, was Bill King, a former submarine commander who had joined the Royal Navy in 1924 at the age of 14. He had been the first man to be catapulted in an airplane off the deck of a ship, and he had seen hard service, mostly beneath the sea, through the whole of World War II. Since then he had lived on his farm in County Galway, Ireland, where he raised black cattle and rode to hounds in top hat and knee boots with the Galway Blazers hunt. He had once raced aboard yachts with friends and sailed his own boat, *Galway Blazer,* across the Atlantic and through the West Indies, yet for the previous eighteen years family life had kept him largely anchored to his farm. But a man who has once been catapulted off the deck of a ship will not slide somnolently into retirement, and Francis Chichester's voyage had fired his imagination.

"It struck me that I could sail alone around the world *without* stopping to refit in Australia." Gripped by his idea, King approached his friend "Blondie" Hasler for help in designing and preparing the ideal boat for such a voyage. Hasler was a former Royal Marine, a World War II hero, and one of the participants of the first OSTAR. He had sailed alone across the Atlantic four times in his own small boat. He agreed to help King and brought in yacht designer Angus Primrose, codesigner of Chichester's *Gypsy Moth IV,* whose design office, Illingworth and Primrose, had turned out a number of famous English racing yachts.

What they came up with was something strikingly unusual: a 42-foot-long schooner with a rounded, turtle-backed deck. The idea behind such construction—the same as for an egg, a bottle, or a submarine—is that there are no potentially weak right-angle joints bolted into the boat, no deck-to-hull or cabin side-to-deck joints, which seas can smash against and weaken or possibly break. It might look odd and humpbacked, but sailors and designers could appreciate and generally agree with the thinking that produced such an appearance. It also offered maximum protection for its captain: The cockpit was below deck, sealed off by two small round hatches, port and starboard, each about the size of the hole in the middle of a kayak;

King could do all sail-handling from the waist up from here without actually climbing out on deck. The whole boat was to be built of thin layers of wood laminated together with glue and wrapped around sectional frames and bulkheads, a sound (and today very popular) wooden boatbuilding technique known as cold-molded construction. The rounded deck and hull would form one integral structure, rather like a plywood tube. This would make for an immensely strong boat.

But this was strength through engineering. Textbook strength. At 4½ tons, King's lightweight, easily driven 42-footer would weigh almost exactly half the weight of Robin Knox-Johnston's slow, fat, 32-foot-long *Suhaili.*

The rig Blondie Hasler drew up for King's boat was a much more radical and iffier proposition for a long voyage through the Southern Ocean. It was an unstayed, junk-rigged schooner: that is, two masts, tapered like flagpoles, stepped on the keel at the bottom of the hull, and secured only where they passed through the deck. No wire rigging would hold the masts in place; their own flexibility was supposed to keep them upright and absorb the loads transmitted by the sails. A single folding Chinese lugsail, easily reefed and handled by one man, would sit on each mast—the same rig as a Chinese junk. Hasler's own boat, *Jester,* a single-masted sloop, was similarly junk-rigged, and the system had by then proven itself on two round-trip Atlantic crossings. But whether masts with no wires to support them would stand up to the winds of the Roaring Forties or a possible capsize could not be known. It was a bold and adventurous choice.

The original estimate for the cost of his boat was £7,000, which King felt he could afford. But the figure quickly rose to £10,000 and in order to defray some of this cost, he looked about for some sponsorship.

It may be hard today, with the five-figure credit card balances that are so readily carried by banks, and the fantastic endorsement remuneration paid to athletes and adventurers, to imagine how difficult it would have been to raise £3,000 in England in the

1960s. Money was far tighter then: food rationing, a hangover from World War II, had still been in force only a decade earlier, and £3,000 would have been a good annual salary in Britain. Watch manufacturers, financial institutions, food and beverage companies were largely unacquainted with the notion of paying a man to go off and risk his life on an adventure in return for some stake in his fame. The rewards were dubious, unquantifiable, and what if he killed himself? Such feats had always been the province of the oddball, someone who hardly embodied a corporate identity. The media—CNN, sports magazines, cable TV programs—that today so completely cover an around-the-world balloon trip, exploiting a Breitling watch sponsor-adventurer connection, simply didn't exist thirty years ago. Coverage was usually a small newspaper report and perhaps a poorly written book. Sponsorship for extreme adventure was in its infancy, and mainly restricted to advances from newspapers—which usually got a good story if the adventurer lived or died—and book publishers. That is where Bill King went looking for his £3,000.

It didn't hurt that he actually resembled Francis Chichester: tall, wiry-thin, bookish, a vegetarian like Chichester, but 8 years younger. With a lifetime at sea behind him, and a sexy boat designed by a famous team, he appeared a serious and credible contender. The *Daily Express* and *Sunday Express* newspapers provided the £3,000 in return for exclusive rights to his story.

Souters, of Cowes, on the Isle of Wight, an old and reputable boatbuilding firm noted for the excellence of its wooden boats, began construction of the yacht, which King named *Galway Blazer II,* before the end of 1967.

In 1966, John Ridgway, a captain in the British Army's Parachute Regiment, and regimental sergeant Chay Blyth rowed across the Atlantic in an open 20-foot-long dory.*

*This was not the first time two men had rowed across the Atlantic: George Harbo and Frank Samuelson, Norwegian immigrant fisher-

The rigors of army parachute training and Arctic survival camps were too controlled for Ridgway. He craved a wilder, more dangerous, less protected experience, but he couldn't imagine what that might be until he heard a radio interview with journalist David Johnstone, who was planning to row across the Atlantic. He immediately got in touch with Johnstone and asked if he could join him. When they met, Johnstone quickly became convinced that he didn't want to spend much time in a small boat with the tough and assertive army captain. Ridgway then decided to mount his own rowing expedition, army-style, and was joined by volunteer Blyth. Though primarily excited by the simple idea of an epic row, Ridgway was well aware that by copying David Johnstone he had created a race between the two boats, and he wanted to be first.

David Johnstone and another journalist, John Hoare, rowed away from the southern end of the Chesapeake Bay on May 21. That placed them close to the eastbound currents of the Gulf Stream, which would be a significant boost across the Atlantic.

Boat preparation and Ridgway's brief spell of blood poisoning delayed the two soldiers. They finally left from Cape Cod on June 4. For weeks, they made dispiritingly slow progress until they reached the strong eastbound current of the Gulf Stream. They found the voyage and the blistering repetitiveness of rowing a monumental bore, but they were good partners, with different, complementary strengths, and they shared a perverse, soldierly satisfaction with their self-imposed hardship. They reached the Aran island of Inishmore off the coast of Galway, Ireland, after 92 days at sea.

Johnstone and Hoare were never seen again. A hurricane crossed their path on September 4. Their waterlogged boat, containing Johnstone's diaries, was found in the middle of the Atlantic on October 14.

men from New Jersey, did it first in an 18-foot lapstrake dory, departing from New York City on June 6, 1896, reaching St Mary's in the Scilly Islands 55 days later.

Ridgway and Blyth might simply have had better luck (a critical but ungovernable factor in all adventures at sea, large and small) than the journalists, but it must have helped that they approached and carried out their epic row like a military mission. Certainly they proved themselves to be tough and determined men.

The two soldiers were treated to a spell of fame. They appeared on television, gave talks everywhere, and were entertained by the queen at a Buckingham Palace cocktail party. Ridgway was thrilled to meet his sovereign, on whom he had once had a worshipful crush, but he felt uneasy in the role of celebrity. Wined and dined, giving slide shows, meeting the rich and famous, he gained almost 30 pounds, and he felt adrift and lost in the social whirl. What he had done, the Atlantic crossing, was, he now found, reward enough in itself. He yearned for more action, something hard and real to take him away from the excess of good food and the glamor that he knew was fleeting and unreal. Chichester's voyage suggested the antidote: "The one thing left untried was a single-handed voyage from Britain right round the world and back to Britain without calling at any port."

Ridgway began thinking about a nonstop solo circumnavigation—but to begin two years later, in 1969. The 29-year-old army officer had some sailing experience, but none of it single-handed, and he wanted first to compete in the 1968 OSTAR race. That race, he felt, would be a good trial run for a circumnavigation, which he would embark on the following year.

In June 1967 he met David Sanders of Westerly Marine, the builder of the Westerly 30, a production fiberglass cruising yacht. A bilge-keeler—it had two shallow fin keels on its bottom instead of a single deep one—the Westerly was a compromise of design that traded speed and stability for access to shallow waters. It was a good boat for small families cruising Britain's coastline who might also want to poke up inland creeks and rivers, where twin keels have the advantage of making the boat rest evenly upright when dried out at low tide. But Sanders was eager to have his boat tested in deep-sea conditions, in which he believed it would perform well. The prestigious OSTAR would provide the perfect sea trial. He lent Ridgway the company's demonstration model to practice sailing alone and offered to underwrite with other sponsors, the cost of a new Westerly 30 for the race if Ridgway found the boat satisfactory.

One of the stipulations made by the OSTAR organizers was that each entrant must have completed a 500-mile nonstop solo passage at sea to qualify. In late July, Ridgway sailed the loaned Westerly from Plymouth, Devon, to the Fastnet Rock off the southern coast of Ireland and back. The 6-day passage went well, giving him a taste of bad weather and the confidence that he could manage the Atlantic alone. When he returned, he told Sanders he was satisfied with the boat and wanted the new Westerly 30, beefed up by its builders, to be ready for the race the following year.

But no amount of strengthening could compensate for the Westerly 30's inherent unsuitability for an Atlantic crossing, let alone, if Ridgway was to pursue his goal of a solo circumnavigation, the tsunami seas of the Southern Ocean. It was the most mistaken of all possible choices available to a man with some sea experience and limitless advice if he sought it. But he had rowed across the Atlantic in a 20-foot dory, and there were few people who would presume to tell such a man what he could not do.

S

2

BERNARD MOITESSIER WAS BORN IN HANOI in 1925, when Vietnam was still called French Indochina. He grew up in Saigon, the priviledged son of a French colonial businessman, but he also learned to speak Vietnamese and absorbed a native sensibility of the East. A result of these two conflicting influences was a yin and yang of worldly ambition and ascetic mysticism that warred in him all his life.

His idyllic childhood world was smashed when the Japanese invaded Vietnam in 1940. Moitessier and his family were briefly interned. After the Japanese pulled out at the end of World War II, he served in a gunboat with French national forces fighting against the Viet Minh communists for control of Indochina. This was the beginning of what became the wider war in Vietnam.

Then, at 27, he sailed away. His homeland overrun with an intensifying war, a brother and close friends dead, Moitessier bought a crude native boat, which he named *Marie Thérèse,* and sailed slowly westward across the Gulf of Siam and the Bay of Bengal. His navigation skill was in its infancy, and he ran aground on the Chagos Bank in the middle of the Indian Ocean, losing his boat. He spent three years in Mauritius, where, with help and donations, he built a ketch, *Marie Thérèse II.* He sailed on as far as the Caribbean, where he shipwrecked a second time, again with the total loss of his new boat.

Brought ashore in Trinidad, Moitessier briefly and desperately considered building a boat out of scrap wood and newspaper painted with pitch, but instead took the advice of the Norwegian consul in Trinidad, an old Cape Horner, who told him, "If you stay in the Antilles you will always be poor. Go to Europe where people are rich." The consul found him a seaman's berth aboard a small tanker, and Moitessier worked his way to Hamburg and finally to France. From these early and inauspicious adventures, he wrote a book called, appropriately, *Vagabond des mers du sud* (the English title more prosaically hinted at his ultimate destinations: *Sailing to the Reefs*).

The book was a best-seller in France. Moitessier was lionized by the yachting community; he got married and passed through that magical and mysterious looking glass that comes with fame and success, on the far side of which people come to you unbidden and ask if they can give you what you want. French yacht designer Jean Knocker offered to draw up the plans for Moitessier's next boat for free. Then, businessman and amateur yachtsman Jean Fricaud offered to build the new boat out of steel in his boiler factory for the cost of the steel plate. Knocker was a designer of repute, and after twice piling his boats up on reefs, Moitessier liked the idea of boilerplate construction.

The new boat was 39 feet long, ketch-rigged, with a long bowsprit. Today it would appear conservative and old-fashioned, but when launched in 1961 it represented the ideal ocean-crossing, live-aboard cruising yacht, for which it would still prove efficient and more than adequate. Moitessier christened his yacht *Joshua*, after Slocum, the old sea captain who had sailed alone around the world. The tough hull was crudely fitted out: telephone poles for masts, phone-company galvanized-wire rigging. But she was as strong as an icebreaker. After two seasons sailing *Joshua* out of Marseille as a sailing school boat, Moitessier and his wife, Françoise, sailed for Polynesia. They crossed the Atlantic to the Caribbean and reached the Pacific through the Panama Canal.

Joshua proved to be an ideal blue-water vessel. Her crude but strong rig, heavy displacement, long keel, and hull shape enabled her to be driven hard, yet easily steered by wind vane, and her motion was sea-kindly and comfortable.

Françoise had three children by a previous marriage who had stayed behind in France at school and with her parents. Bernard and Françoise had told the kids they'd return as soon as possible, but from Polynesia this meant continuing on westward around the world by normal trade wind cruising routes, another year of sailing before reaching France by way of the Red Sea and the Suez Canal. By the time they reached Tahiti, however, Françoise was missing her children badly and Moitessier had come up with another, much faster route home. He proposed sailing *Joshua* south to the Southern Ocean, turning east along the clipper ship route, and hurtling home to Europe, nonstop, though the Roaring Forties by way of Cape Horn. It would take four months instead of a year. Provided a yacht was strong and sound, which they both believed *Joshua* to be, it was, Moitessier told his wife, a logical and reasonable route. Although he warned her of the conditions they could encounter, Françoise could have had no idea what they were facing. Nor, as it turned out, had Moitessier.

They sailed from Tahiti on November 23, 1965. It was a voyage that would make small-boat history. On December 13, at 44 degrees south, in the loneliest wastes of the Southern Ocean, the barometer began a steep fall and the wind started to blow hard from the northwest. By 6 A.M. the next morning, they were running downwind under bare poles (all sail stowed) before a whole gale—winds of 40 to 50 knots. The barometer was still falling, the wind still rising. The wind vane steering gear had become overpowered, no longer able to keep *Joshua*'s stern to the wind and sea, so it had been disconnected and the vane stored below; steering was now by hand, and would continue so as long as the storm lasted.

Moitessier knew, he thought, exactly what to do under such conditions. Before leaving Tahiti he had met and spoken with American sailor William Albert Robinson, who had run his 70-

foot brigantine *Varua* under bare poles through these same seas, in what he had described as the "ultimate storm" in his book *To the Great Southern Sea.* Robinson had followed the accepted storm technique of dragging lines towed astern with anchors or ballast attached to them to slow the yacht down as it runs before wind and seas and prevent it from going so fast that it becomes unmanageable and broaches, possibly capsizing. The technique had worked for Robinson. Moitessier had read Robinson's book, and another, *Once Is Enough,* Miles Smeeton's thrilling (and even funny) account of his two separate and disastrous capsizes aboard his yacht *Tzu Hang* in the Southern Ocean west of Cape Horn. All small-boat sailors heading south into these seas (and many enthralled armchair adventurers) have read these two classics, profoundly hoping not to meet the same conditions and to learn what might possibly help them if they do. Smeeton too had towed lines, but that hadn't prevented *Tzu Hang* from being hurtled forward to bury her nose in the sea when her stern was lifted by a giant wave that pitchpoled the yacht, snapping both masts off clean and ripping off the entire deckhouse, leaving a gaping hole in the yacht's deck. Smeeton wondered, in his book, whether any precautions or techniques could have made a difference to the great sea that flipped *Tzu Hang,* but a sailor was conscientiously bound to follow procedure. (Under jury-rig, with a short mast made from floorboards, *Tzu Hang* limped into Valparaiso, where Smeeton and his singularly adventurous wife, Beryl, spent the better part of a year repairing their yacht. They sailed south for the Horn and *again* were rolled over in a storm, suffering the same damage.)

Moitessier had prepared; he was all ready to slow *Joshua* down. He soon had five thick lines, between 16 and 55 fathoms (96 and 330 feet) long, trailing astern. Three lines were weighted by two or three 40-pound lumps of pig iron. A fourth line dragged a large net as a sea anchor, and the fifth trailed freely with nothing attached to it. This tremendous drag certainly slowed *Joshua*'s progress; breaking seas now swept over the decks and the boat appeared to be standing still or going backward.

Moitessier was steering from a small wheel inside the boat, beneath a turretlike hatch he had cobbled together in Tahiti from a steel washbasin and Perspex windows, which allowed him to look out but stay dry. He and Françoise had taken turns steering, but the lines astern had made the boat sluggish to respond and now only with difficulty and intense concentration could he alone keep the boat on course—the "course," in this extreme, being to keep *Joshua*'s pointed stern, its least vulnerable aspect, dead before the enormous breaking waves. They could no longer move around inside the boat except by crawling and gripping onto handholds. Moitessier clung to the wheel at his perch, Françoise burrowed into their bunk. Food, except for what was easily grabbed in one hand, was out of the question.

Another entire day and night wore by and morning came again; the barometer continued to drop to a rare low and the storm intensified. The waves grew nightmarishly large, as high between trough and crest as eight-story buildings, sweeping faster over the sea and carrying the boat unstoppably with them, until Moitessier believed that despite all his preparation and reading and techniques, *Joshua* was on the point of being overwhelmed.

The awful truth now hit him: *Joshua* might be the perfect cruising boat for trade wind seas, but she was fatally out of place in the Southern Ocean, in the storm in which she now found herself. Disaster, inescapable, loomed.

But Moitessier's mind recoiled at the conclusion that *Joshua* could not make it where other boats—good boats but no better, he finally believed, than his own—had come through. At this desperate moment, when he felt his boat to be on the point of foundering, he thought of another boat, another book, another sailor: the Argentine Vito Dumas, who had sailed alone around the world in 1942–1943 aboard a double-ended ketch, *Lehg II*, a yacht with a shape very like *Joshua*'s, but at 31 feet long, considerably smaller. Dumas's book, *Alone Through the Roaring Forties*, is another in the pantheon of must-read classics that deal with Southern Ocean sailing, and Moitessier remembered that Dumas claimed to have carried at least a small staysail while running before the wind in *all* weather—in the worst of weathers—clearly therefore running at speed, *not* slowing down, in conditions such as these.

Then a wave caught *Joshua,* not directly astern, but partly slewed around at an angle, and despite all the lines and weight dragging in the water, she was carried forward at fantastic speed. Yet instead of plunging down and burying her bows in the wave's trough, the wind heeled *Joshua* over on her side, so that she planed like a water ski along the surface of the breaking wave. The wave passed harmlessly beneath the boat, and Moitessier had discovered Vito Dumas's secret.

"Quick!" he shouted to Françoise. "Take the helm for two seconds."

He grabbed his Opinel, the little wooden-handled French pocketknife with a steel blade that keeps a wonderful edge, climbed out on deck, and quickly cut away all five trailing lines.

Back at the helm, he immediately noticed the change. Gone was *Joshua*'s sluggishness. No longer a sitting duck to be pounded and swept by the great seas, she now raced away before them. He ran the boat downwind as before, but as each wave approached, he gave the wheel a slight turn at the last minute and took the wave at an angle of 15 to 20 degrees. The wind hit her side, heeled her over, and off she flew, planing across the surface of each wave. The speed gave her rudder greater control and she responded instantly to the helm when the wave was past as Moitessier brought her stern into the wind again. The enormous waves, their apparent force reduced by *Joshua*'s speeding away from them, now rolled harmlessly beneath her quarter.

The storm lasted 6 days and 6 nights. Bernard Moitessier steered through more heavy weather, and learned more about handling it, in those 6 days than most sailors do in a lifetime: a compression of experience that turned him into a master mariner in a single week; a man who had spent a short eternity at the farthest reach of all sailors' fears.

Four months later, the Moitessiers dropped anchor in Alicante, Spain, their first stop, 14,216 miles from Tahiti. Without intending to, trying simply to get home fast because they missed the kids, they had made the longest nonstop voyage in a yacht to date—a world record, and by way of the dreaded Horn.

Moitessier very quickly wrote another book, his second, about their voyage, *Cap Horn à la voile* (titled in English: *Cape Horn: The Logical Route*), which was published in time for France's premier boat show, the Salon Nautique. It became a huge best-seller. In France, where long-distance sailors enjoy the sort of movie star celebrity known only to sports figures like Michael Jordan today in the United States, Moitessier became a national hero. Awards were heaped on him. In England, the "Moitessier method" was discussed at a *Yachting World* forum on heavy-weather tactics, where it was pronounced "rather startling." By the end of 1966 he was world-famous.

And very unhappy. He felt he had dashed off his book too fast, rushing it to coincide with and augment the glory of the Salon Nautique. He felt, he wrote later, that in so doing he had committed a crime. Moitessier experienced nothing in moderation. His books are written with an ingenuous, exuberant lack of restraint (and editing), full of a sensual exultation of sea. When he was up, he was way up. But when he was down . . .

> October ('67) was devastating. Wrapped in total silence, sucked down by a huge inner emptiness, I sank into the abyss. . . . I felt madness burrowing into my guts like some hideous beast. I found myself wondering what last thoughts come to someone who has swallowed a lethal dose of poison.

This goes beyond remorse for skimping on a book. It seems more likely that following his epic voyage and starburst of fame and glory, he was experiencing a pronounced bipolar slump.

He got himself out of it with a typically intense swing back into the stratosphere.

> I must have been on the point of suicide when . . . in one blinding flash . . . I saw how I could redeem myself. Since I had been a traitor by knocking off my book, what I had to do was write another one to erase the first and lift the curse weighing on my soul.
>
> A fresh, brand-new book about a new journey . . . a gigantic passage. . . .
>
> Drunk with joy, full of life, I was flying among the stars now. Together, my heart and hands held the only solution, and it was so luminous, so obvious, so enormous, too, that it became transcendent: a non-stop sail around the world! . . . And this time I was setting out for the battle of my life alone.

He doesn't mention Chichester. But his blinding flash seems to have come around the end of 1967. Chichester had sailed home that May.

S

3

WINTER IS A DISMAL SEASON for sailors in England. The weather offshore is bad, even dangerous, and the misty coastal glim has become impenetrable and uninviting. Boats are cold and damp and under wraps in their mud berths or puddled boatyards. But in January the gloom is pierced by the arrival of the annual London Boat Show. Boaters, boatbuilders, yacht designers, ship chandlers, brokers, and sailing journalists from all over the country come in out of the rain and look at new boats and hardware. They finger the new rain gear, talk about anchor chain, and try to find some aspect of it all that might appeal to their wives and girlfriends. And they pore through cruising guides to Brittany, or the Bahamas, or anywhere a boat can sail to. So it is also the season of their dreams.

One of these dreams, that January 1968, was epic, Ulyssean, and talk of it passed through the yachting community like contagion. Everybody now knew what the next great voyage must be: a nonstop solo circumnavigation. So when it was rumored at the boat show that Bernard Moitessier was preparing for another voyage, the shape of it was easily guessed at. The details of Bill King's *Galway Blazer II*, then under construction, were released to the newspapers during the boat show, and his intentions became known to all. And there was talk of others making similar plans.

John Ridgway, reading of Bill King's proposed voyage, made a "military appreciation of the situation" and decided to advance his plans by a year. He would forego the transatlantic race, which he had looked on as good experience for a circumnavigation, and concentrate wholly on the bigger voyage. His literary agent immediately contacted *The People* newspaper, which had sponsored his transatlantic row, and they agreed to back his circumnavigation. In order to get a jump on King and his bigger, faster boat, which wouldn't be completed until July or August, Ridgway determined to depart on June 1. This date would mean reaching the Southern Ocean sometime in September, rather too early for comfort, at the beginning of the southern spring, but would put him off Cape Horn in January—midsummer—the safest time for a passage south of the stormy cape.

The *Sunday Times,* Britain's upmarket Sunday newspaper of record, which had sponsored Francis Chichester and reaped a bonanza with that story, was very keen to get itself linked with one of the nonstop circumnavigators. The paper dispatched Murray Sayle, the reporter who had covered the Chichester story, to assess the growing pool of possible contenders for this last great sailing "first."

He liked "Tahiti Bill" Howell, a 42-year-old Australian who had spent years sailing through the South Pacific supporting himself as a cruising dentist. Howell came with an admirable resumé: he had sailed the 24-foot *Wanderer II,* a famous yacht once belonging to the venerable English sailor Eric Hiscock, from England to Tahiti with one crew, and from there single-handed to British Columbia by way of Hawaii. He had sailed a 30-footer to sixth place in the 1964 OSTAR. Now he had a 40-foot catamaran, *Golden Cockerel,* capable of sailing much faster than any monohull. He was planning on racing his cat across the Atlantic in that summer's third OSTAR, then immediately turning left once he was across the finish line and heading south for his nonstop circumnavigation.

If it was to sponsor anyone at all, the *Sunday Times* had to act fast. King had received sponsorship from the *Sunday Express,* Ridgway had *The People.* Both men, sailing new boats, seemed like possible winners. If the *Sunday Times* didn't jump on "Tahiti Bill," it could find itself rooting among the dreamers with inferior boats and untested fortitude.

But who were the others—the unknowns, the undeclared, the dark horses that would appear and steal the show? Who would sponsor them?

Murray Sayle and Ron Hall, Sayle's department head at the *Sunday Times,* hit upon the idea of sponsoring a race that would include everybody. It occurred to them simultaneously but independent of each other, and initially each had different ideas for such a race. Murray Sayle correctly believed that what would matter most, to the competitors and to the public, was who would circumnavigate alone and nonstop first. That was the sailor the history books would remember.

Ron Hall thought that if there was to be a race, the sailors would have to compete in some way that would give them all an equal chance. But an official start, with everyone setting off at the boom of a gun, was out of the question. The men now anxiously preparing their boats, all of them increasingly aware of the efforts of a growing number of rivals, would certainly leave the moment they were ready, if not before. Some of them had already made arrangements with publishers and other newspapers. If the *Sunday Times* proposed a race, the sailors might refuse to enter. A race was in fact the last thing any of them wanted. These were not yachtsmen or sportsmen. They were hardcase egomaniacs driven by complex desires and vainglory to attempt an extreme, life-threatening endeavor. Each had powerfully visualized what must be done, and was consumed with the need to do it first. They were loners. No one would be waiting for anybody else.

Yet the race to be first had already begun.

One afternoon in March, Murray Sayle and Ron Hall sat down together and came up with an ingenious way to scoop all their newspaper rivals and put the inevitable race firmly in the hands of the *Sunday Times*. The newspaper would offer a trophy, which the two journalists decided then and there to call the Golden Globe, for the first sailor home. An additional award, answering Ron Hall's wish for some sporting measure, would be a cash prize of £5,000—a princely sum in 1968—for the fastest voyage. Thus, if the first man home won the Golden Globe simply because he had been the first to set out, there would still be an incentive for others to race home for the fastest time and the cash prize. Contestants were not even required to enter the race. Anybody setting out, sponsored or not, whose departure and arrival dates could be verified, would be eligible for the *Sunday Times* prizes. This way, no circumnavigator could *not* take part.

The rules were simple and designed to embrace the various plans already afoot: competitors could leave from any port of their choice in the British Isles, on any date between June 1 and October 31, 1968. (To leave earlier or later could mean reaching the Southern Ocean and Cape Horn during the severer weather of the austral winter, and—in what would turn out to be its only such effort—the *Sunday Times* wanted to avoid encouraging undue risk.)

The route was around the world "by way of the three capes," in clipper ship parlance: the Cape of Good Hope in South Africa; Cape Leeuwin, Australia; and South America's notorious Cape Horn. Competitors would sail alone, without stopping or putting in at any port, without assistance or resupply. And they would return to their port of origin.

To give luster and authority to its self-appointed interest, the *Sunday Times* quickly put together a panel of judges, made up of heavyweights from the world of yachting, with Sir Francis Chichester as its chairman.

With plans for a circumnavigating yacht drawn up for him by Colin Mudie, Robin Knox-Johnston went looking for a quote from a builder. The design, 53 feet long, simple and lightweight,

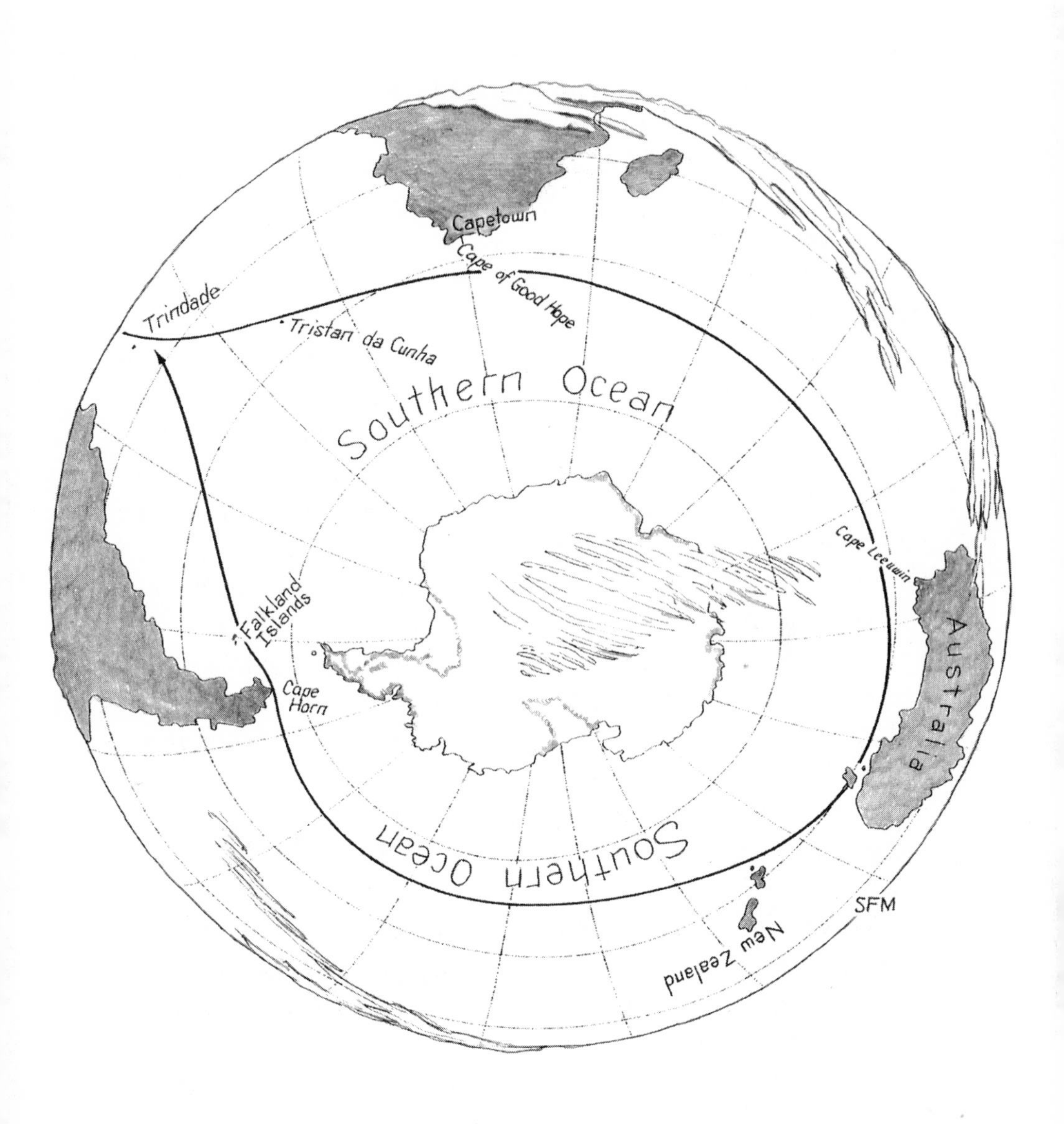

Around the World by Way of the Three Capes

though to be built of steel, was of such unusual construction that most boatyards either refused to quote at all or put him off with outlandishly high figures. He finally found a barge-building yard on the River Thames that quoted £2,800 for the hull, and even appeared interested in the project. This price meant the possibility of a complete boat ready to sail for around £5,000, which was cheap.

But Knox-Johnston had no money at all. By the end of 1967, he had written over fifty letters to firms and businesses seeking sponsorship, and those who replied had all declined. *Suhaili,* his only asset, was still up for sale, but roughly built and old-fashioned, she was few people's idea of a proper yacht, and there had been no takers.

Yet he had made up his mind to go. Increasingly his thoughts turned to *Suhaili,* the boat he had sailed around the Cape of Good Hope and across 10,000 miles of ocean. Between professional stints at sea, he had lived aboard her for two years. He knew her, she was a proven vessel—and she was available.

She would need a refit, new sails, new rigging, some sort of self-steering gear, and he would have to load her with a year's supply of food—but all this seemed possible. He turned his back firmly on the sleeker pipe dream and decided to go ahead with *Suhaili.* He would find the money for what he needed where he could.

George Greenfield, a London literary agent, thought there might be a book in Knox-Johnston's voyage. Greenfield specialized in adventurers—the newly knighted Sir Francis Chichester was one of his clients, as was British explorer Wally Herbert, who was then putting together the British transarctic expedition, a sledge-hauling crossing of the Arctic icecap by way of the North Pole. Greenfield thought he saw the same stuff in the young merchant seaman. He might be unknown in yachting circles, but he had sailed his boat home from India, and Greenfield was excited about the idea of his nonstop circumnavigation. He told Knox-Johnston to get on with his preparations.

Early in 1968 Greenfield sold the book rights to Knox-

Johnston's voyage to the London publisher Cassell. This brought an advance payment of enough money for *Suhaili*'s refit. Then he approached the *Sunday Times,* hoping for the same coverage he had got for Chichester, but the paper was unimpressed—it was the decidedly poor prospect of Knox-Johnston and his archaic Indian-built boat that led the *Sunday Times* to back Tahiti Bill Howell. Finally, Greenfield got the London *Sunday Mirror* newspaper interested in buying the rights to exclusive accounts radioed from Knox-Johnston at sea. Before signing, the *Sunday Mirror* people wanted to meet their sailor, and Greenfield set up a nautically themed lunch on a restaurant boat on the Thames. During the lunch, a tug went by on the river throwing up a wake that rocked the restaurant, and the tough young adventurer who was going to sail alone around the world lost his balance and fell out of his chair. Greenfield still made the deal.

With enough money for a full refit and all his supplies, Knox-Johnston now devoted all his free time—between Royal Navy reserve duty aboard HMS *Duncan,* for which he was already committed—to preparations.

Bernard Moitessier had already signed the contract with his French publisher, Jacques Arthaud, for the book he would write of his epic nonstop, single-handed voyage. He spent the spring of 1968 preparing *Joshua* in the French Mediterranean port of Toulon. When the *Sunday Times* learned of his intentions, Murray Sayle was sent to invite him to take part in its race. He found the Frenchman in a portside bistro.

Moitessier was aghast. He had made a pact with the gods (he later wrote) to make up for selling his earlier book down the river and felt that his motives for the new voyage must remain as pure as driven snow. He told Sayle that the idea of a race made him want to vomit. Such a voyage, he said, belonged to a sacred domain where the spirit of the sea had to be respected. A race for money and a gold-colored ball would make a circus of all their efforts. He got up in a rage and left the bistro.

Sayle and his newspaper were flummoxed. Moitessier and his boilerplate-steel boat unquestionably represented the strongest effort underway. The mystic Frenchman and his *Joshua* were the only sailor and boat of those now preparing to have already spent any time in the Roaring Forties. Together they had been around the Horn, the great, fearful, spectral bogey of the Southern Ocean. Moitessier could sail around the world from Toulon to Toulon and make a mockery of the Golden Globe race. Unwilling to lose its most formidable competitor and see its race made redundant, the *Sunday Times* revised a rule for the Golden Globe trophy for first man home. It made it eligible to anyone starting from any port north of latitude 40 degrees north—that's just below the French-Spanish border.

A few days later, Sayle cornered Moitessier again. He started by suggesting (according to Moitessier) that his famous Tahiti-Alicante voyage had been the catalyst for Chichester's voyage (unlikely, since Chichester's preparation for his circumnavigation had begun well before *Joshua* reached Alicante), and for all the voyages now being planned. But Moitessier didn't need any sweet-talking; he had already decided to join the race. He had thought of nothing else since Murray Sayle had first spoken to him. He told the reporter that he'd join his race, and would sail *Joshua* to Plymouth, Devon, start from there and return to Plymouth to be eligible for both prizes. And if he came home first and fastest, he'd snatch the check without a word of thanks, auction off the Golden Globe, and thereby show his contempt for the *Sunday Times*.

This was pure Moitessier, the yin and yang of his childhood influences battling within him. The pleasure and the unease he would feel all his life about the fame and money that came his way, usually from simply doing what he wanted. His ambivalence about the race certainly had much to do with his feeling that it would sully the purity of his intent and effort. But he realized that it would be a great and historic race, with worthy rivals, and that if he didn't take part, he would be left out. He also knew he stood a very good chance of winning both prizes and seeing his star blaze brighter—and he couldn't resist that.

On March 17, 1968, the *Sunday Times* announced its Golden Globe round-the-world race.

> **£5,000**
>
> The £5,000 *Sunday Times* round-the-world race prize will be awarded to the single-handed yachtsman who completes the *fastest* non-stop circumnavigation of the world departing after June 1 and before October 31, 1968, from a port on the British mainland, and rounding the three capes (Good Hope, Leeuwin and Horn).
>
> **GOLDEN GLOBE**
>
> The *Sunday Times* Golden Globe will be awarded to the first non-stop single-handed circumnavigator of the world. The yacht must start and finish at the same port in a northern latitude (north of 40 degrees N) and must be round the three capes.
>
> In either case, the circumnavigation must be completed *without outside physical assistance* and no fuel, food, water or equipment may be taken aboard after the start.
>
> Those are the only conditions. The same yachtsman may win both awards. No formal entry is necessary. . . . Yachtsmen of any nationality are eligible and yachts may be of any design and built anywhere. Any sponsorship is permitted and the *Sunday Times* asks nothing of the winners. . . .
>
> Sir Francis Chichester, chairman of the Committee of Judges, has . . . expressed concern about the hazards that lie ahead on such a voyage. It cannot be stressed too often that while, in theory, any person may take part in this event . . . only men with considerable long distance ocean-yachting experience should consider competing.

Elsewhere in the paper, Chichester, the paterfamilias seadog, was quoted:

> Some of these chaps don't know what they are letting themselves in for. If any of them succeed in getting round it will be remarkable. By comparison the Atlantic is about on the level of a canoe trip across the Serpentine.

The list of competitors was impressive: Bill King, the former submarine ace, and his revolutionary *Galway Blazer II;* John Ridgway, the supremely fit army captain who had already rowed across the Atlantic; Tahiti Bill Howell, the sailing dentist, 20,000 miles of Pacific cruising in his wake, most of it single-handed, making him the most experienced single-hander of all the competitors, sailing potentially the fastest boat; Robin Knox-Johnston, the merchant navy captain, about whom little else could be said; and Bernard Moitessier, the famous French yachtsman, known for his epic Tahiti-Alicante voyage in his yacht *Joshua* and (reported the *Times*) for his dry sense of humor.

Another sailor who had caught the bug of the great voyage was Donald Crowhurst, an electronics engineer and manufacturer of the Navicator, a radio navigation device designed for the yachting market, which he was selling from a booth that January at the boat show. He had no boat of his own, so he tried to convince the Cutty Sark Society, which was planning to purchase Chichester's *Gypsy Moth IV* (and mount it beside the square-rigged sailing ship *Cutty Sark* near the National Maritime Museum in Greenwich), that he, Crowhurst, should sail it around the world again, nonstop.

Crowhurst appeared to be an experienced yachtsman. He had impressed a number of people, including Angus Primrose (the codesigner of *Gypsy Moth IV* and now *Galway Blazer II*) with his arguments for the loan or charter of *Gypsy Moth IV.*

At the time, there was considerable public and yachting community antipathy to the idea of retiring such a famous yacht to a permanent cement berth. Many felt *Gypsy Moth IV* should be kept afloat and used for precisely the sort of noble and epic voyage Donald Crowhurst was proposing. The Cutty Sark Society eventually refused him, and *Gypsy Moth IV* became a soot-covered museum display. But Crowhurst was not put off.

Four days after the *Sunday Times* announcement, Donald Crowhurst declared himself a competitor in the race. Few paid attention. In the following six months he was mentioned only once by the *Sunday Times* in a roundup of possible competitors. No information was given about him, except that he planned to sail in September and that his boat was a "revolutionary ketch."

Here was the dark horse. Here was the story that would steal the show.

4

In 1968, Donald Crowhurst was 36 years old. This is an age when most men—that is, business and professional men, not artists or bohemians or mystics, but men with families, responsibilities, and middle-class aspirations—are beginning to have some success in their work, if they are ever to have any at all. And it's an age when most people are beginning to see themselves clearly, whether they like what they see or not.

In a modest way, Donald Crowhurst had been doing well: After an uneven career, he appeared to be on the verge of success as an electronics engineer, and by the mid 1960s, his company had achieved a flurry of profitability. He was a happy family man with a wife and four young children and a large rambling house in a pretty country town in Somerset. He was a popular, charismatic figure in his community, involved with the local amateur dramatics society, and he'd been voted a councilman for his hometown of Bridgwater.

But Crowhurst's early life had been marked by setbacks and crushed aspirations, so when his business took a downward slide in the late 1960s, and his visions of solid success began to waver and vanish, he felt with awful familiarity the creeping pinch of failure. He looked around, and then into himself, to find something that would stop his fall.

He was born in India, the son of a superintendent with the North Western India Railway Company; his mother was a schoolteacher. Their comfortable raj lifestyle (complete with servants and pretensions unsustainable elsewhere) came to an abrupt end in 1947, the year of India's independence from British rule. The Crowhursts returned to England to find themselves, like so many English colonials when they returned to the "home" that had never been home, in severely reduced circumstances.

Donald's father, John Crowhurst, had put all his retirement savings in a sporting-goods factory in Pakistan which was burned down in a riot soon after they reached England. With what money they had left the Crowhursts bought a small house at Tilehurst near Reading, and then John went looking for work. The only job he could find was as a porter in a jam factory. Donald's mother, Alice, a highly strung, hypochondriacal woman, was appalled by their debasement and anxious at the sudden burdens placed on her husband.

She was right to worry. The following year, John Crowhurst died of a heart attack while digging in his garden. Alice was left with the house, but little money. She found it necessary to borrow from relatives; this brought her a few pounds and much humiliation. She became filled with a bitter sense of superiority blighted by circumstance. She passed a heavy dose of this on to her son.

At the age of 16, Donald Crowhurst was forced to leave school. He had hoped to go to Cambridge University but instead became an apprentice in electronic engineering at a technical college. In England's then rigidly stratified social structure, he keenly felt his family's series of falls from an early position of privilege and his own dashed hopes of a good education and the elevated status that would have come with it. He spoke with the accent of Britain's minority but then-dominant "upper middle class"—an accent that has all but disappeared now, except in satire—and he was not prepared by temperament or upbringing to settle for a station that he felt was beneath him.

At 21, he embarked on a series of false career starts. He joined the Royal Air Force, learned to fly, and assumed the social role of a young officer, tearing about in a sports car, drinking too much, inciting others to wild times. But he got a reputation as a trouble-maker and was dismissed from the RAF. He joined the army, specializing in electronic equipment, and continued his performance as the racy officer. But again he was constantly in trouble: he smashed up his car, was caught trying to hot-wire another car to get back to his barracks after a late night out, and was finally asked to leave the army. He found a job with an electronics firm, crashed the company car, was reprimanded, and quit. Meanwhile, he got married, continued to jump from job to job, moving his growing family around southern England. They came to rest in Bridgwater, in Somerset, where Crowhurst landed a job as chief design engineer with another electronics firm. It was not far from the Bristol Channel. Donald's wife, Clare, was a Catholic, and children began to arrive at regular intervals. Money was always tight; nevertheless, about this time Crowhurst bought himself a small 20-foot sloop and started sailing.

The only constant, apart from his inability to work well for others, was his interest in electronics. At home, he would spend hours in his workshop, bent over wires and circuitry, attempting to invent clever devices. Combining this with his growing interest in sailing, he eventually produced a radio-direction-finding instrument (RDF).

Navigating a boat using the RDF was becoming popular and widespread in the 1960s. An RDF instrument is tuned like a radio (essentially, it is just a radio with a compass mounted on it) to the frequency of a known signal at a known location ashore, then swung by hand until the signal direction is determined (actually by finding the null, or least evidence of the signal), and a compass bearing taken. Two such bearings, from different stations, can provide a fairly accurate offshore position. To the sailor wrapped in fog, foul weather, or darkness, many hours past the last visual fix on a sea buoy or a piece of coastline, RDF added a considerable measure of safety and relief. Like many other navigational devices and systems, it has now largely been supplanted by the ease and fantastic accuracy of the satellite-enabled global positioning system (GPS). But in its day, RDF, along with other radiowave navigational systems, like Decca and Loran, represented the greatest leap in navigation since accurate timepieces made possible the determination of longitude.

Yachting has long been ruefully likened to standing in a cold shower while tearing up money. It's a famously uncomfortable and expensive recreation. But RDF devices, when they appeared, were hardly more expensive than small transistor radios, and soon became standard equipment aboard yachts—particularly those sailing in English waters, where any uncertainty of position is compounded by strong tides and a coastline that looks much the same in one place as another and is easily lost sight of in habitually murky weather.

Crowhurst called his RDF a Navicator. It was not innovative in any way, but it was well-designed and conveniently shaped like a pistol. With its little compass mounted on top, it looked like a ray gun from a 1950s space movie. The user could simply and easily aim and take a bearing.

At about this time, Alice Crowhurst, Donald's mother, swallowed a fistful of sleeping pills in an apparent suicide attempt. She was taken to a hospital and thereafter remained in institutional care. Her house was sold and Crowhurst decided to use some of the money to start his own electronics business. He called the company Electron Utilisation. It was launched with the manufacture of his Navicator.

It was the right product at the right time. Crowhurst soon had six full-time workers in a small factory. Pye Radio, then a household name in Britain as a manufacturer of radios, televisions, and gramophone players, began negotiations with Crowhurst to buy the Navicator. They paid him £8,500, a sum that would have felt like fifty times that amount today. He bought a house, Woodlands, in Bridgwater, and turned its small stable into his workshop. Donald Crowhurst had arrived.

Crowhurst was well aware of the larger world, the bigger ideas, the brighter people beyond his immediate provincial horizons deep in the English countryside. He was more intelligent than most of his acquaintances, and they knew it. "I did not worship Donald Crowhurst," said a friend. "I *recognized* him—as the most vivid and real person I have ever met."

He carried people away with his brilliance. With no one to match him intellectually or egotistically, no one to shoot him down, deflate him, or burst the bubbles he blew, he carried himself away too. He joined the local amateur dramatic society and became one of its stars, but the wider world stage on which he wished to strut seemed out of reach.

His Navicator, his endless tinkering with wires and transistors in search of other inventions, his interest in amateur theatricals, his dominating, supercharged personality, were all symptoms of his great urge to leave his mark. Crowhurst believed he had something important to give the world, and he was constantly trying to find it.

On Sunday, May 28, 1967, as Francis Chichester approached Plymouth—which was only 70 miles from Bridgwater—Donald Crowhurst spent the day sailing with a friend. But they headed out into the Bristol Channel, the other side of England's southwestern peninsula, far from the waters that filled throughout the day with boats waiting for the first sight of *Gypsy Moth IV.* Crowhurst admired Chichester. He had read his earlier books about his lone transatlantic crossings and had closely followed his voyage around the world. Yet that day he turned his back, becoming aloof and scornful. As the two men listened to the BBC's coverage over their boat's radio, Crowhurst derided Chichester's accomplishment. Plenty of people had sailed alone around the world, he said, and Chichester had stopped for a long rest in Australia.

Crowhurst then told his friend that for years he had thought about sailing around the world alone and nonstop. *That* would be something worth making a fuss about.

Later that day, after they returned to port, they went home and, like everybody else, watched Chichester's arrival on television.

Crowhurst's role as a successful businessman was short-lived. Pye Radio backed out of the Navicator deal. Their initial payment gave Crowhurst and Electron Utilisation the appearance of prosperity for a while, but he was eventually forced to abandon his small factory and cut his workforce from six to one part-time assembler in his stable-workshop. The Navicator was not, as he had hoped, to be widely distributed to every ship chandlery in Britain. He was reduced to hawking it from a booth at boat shows.

But his self-belief, his intelligence, his ideas, and his charm were persuasive. Looking for new backers, he was introduced to Stanley Best, a buinessman from nearby Taunton, Somerset, who had become wealthy selling caravans. In 1967 Best made a first tentative investment in Electron Utilisation. It was in the form of a loan of £1,000. Crowhurst's undaunted and enlarging vision held Stanley Best in thrall long past the point where his pragmatic business sense should have stopped him.

"I always considered Donald Crowhurst an absolutely brilliant innovator," Best said later, "but as a businessman . . . he was hopeless. He seemed to have this capacity to convince himself that everything was going to be wonderful, and hopeless situations were only temporary setbacks. This enthusiasm, I admit, was infectious. But as I now realize, it was the product of that kind of overimaginative mind that was always dreaming reality into the state it wanted it to be."

5

JOHN RIDGWAY SAILED from remote Inishmore, one of the Aran Islands, 40 miles off the Galway coast of northwest Ireland, at 11.38 A.M., Saturday, June 1, 1968, the first Golden Globe sailor to depart.

It was a hell of a place to take off from. To get there, Ridgway had sailed his little 30-footer almost 1,000 sea miles from its builder's yard in Portsmouth, southern England, down the English Channel, across the Irish Sea, up the wild western coast of Ireland, taking himself far to the north of the departure points of all his rivals, and in the wrong direction for his route down the Atlantic.

He and Chay Blyth had made landfall on Inishmore at the end of their epic row across the Atlantic almost two years earlier, and Ridgway told the reporters who had gathered to record his departure that he felt an affinity for the local islanders, "who live so closely with the sea, good people, who know what suffering is." The locals reciprocated his feelings. They had erected a plaque in Kilronan Harbour to mark the spot where he and Blyth had come ashore, and danced jigs and reels in his honor at the local parish hall the night before he sailed. But Ridgway had not come all this way to make Inishmore his port

Sailing Route Down the Atlantic Ocean

of departure entirely for sentimental reasons. This was a talismanic choice, its inconvenience and geography strongly at odds with the hard pragmatism of his army training, and in making it he proved himself as classically superstitious as all those seamen who will not begin a voyage on a Friday, or who stab their knives into wooden masts for wind when becalmed. He had come all this way for luck.

But minutes after leaving his mooring he found bad luck at sea. A BBC camera boat, trying to outmaneuver its rival ITN boat, came too close and hit the stern of *English Rose IV.* Ridgway, already anxious and struggling for composure in front of cameras and well-wishers, lost control and screamed abuse at the BBC boat. There appeared to be no damage.

About 20 minutes later, the ITN boat, a chartered 25-ton trawler carrying his 23-year-old wife, Marie Christine, and a boatload of press and well-wishers, ranged up alongside *English Rose IV,* and the sea swell threw the two boats together, making a heavy bump felt by those aboard the trawler. This time, Ridgway was so upset he couldn't talk. But his silence was admiringly noted by newsmen on the trawler, who took it as an indication of considerable aplomb, a glimpse of the tough silent stoicism with which he would meet graver conditons later on.

This collision split *English Rose IV*'s wooden rubrail near the rigging, superficial and mainly cosmetic damage, but it left Ridgway badly unnerved: "I looked down at the splintered strip of wood . . . defeat filled my mind."

Ridgway finally left the pestering boats astern as he sailed out into the empty Atlantic. His voyage had begun inauspiciously, and he couldn't forget it.

Chay Blyth, John Ridgway's partner on the row across the Atlantic, was a man in whom the Ulysses factor coursed thick and strong. He was the perfect example of the way this factor excites and stimulates others who wish to see such a man rise to his singular calling, who gather around him and urge him on to do something none of them would dream of doing themselves.

Much shorter than his 6-foot-tall captain, stocky, and inferior in rank, Chay Blyth was no less an adventurer. He was mentally tougher, and far less given to doubts and introspection than Ridgway. Years earlier, paired as a team, the two had won an arduous 75-mile overnight army canoe race only because of Chay Blyth's unfaltering determination. After capsizing in frigid water and being flushed through the white water of a lock and almost drowning, emerging only to face the quick onset of hypothermia, Ridgway suggested they give up. "No! We're going to win," said Blyth. He pushed them on and they did win. It was a revelation to Ridgway, the degree to which Blyth's absolute mindset had so altered their apparent situation. Before their transatlantic row, the two had shared survival training in Middle Eastern deserts and the Canadian Arctic. Ridgway always acquitted himself well, but Blyth positively embraced hardship.

Blyth had watched Ridgway first prepare for the OSTAR and then set his sights on a nonstop circumnavigation with irresistible envy. It seemed to him the grandest survival test of all. Ridgway had done some sailing and had made his 500-mile solo passage to Fastnet Rock and back, but Blyth had never sailed a mile. It didn't faze him; before he set off to row across the Atlantic, Chay Blyth had never even been aboard a boat. When he decided that he too wanted to sail alone around the world, nobody thought to ask a man who had rowed across the Atlantic what his qualifications were. If Ridgway could do it, so could Blyth.

Exactly as happened with Ridgway, Blyth was readily offered the use of a boat by a company eager to see its product tested in the high-stakes arena of a round-the-world race. The boat, named *Dytiscus III,* was a Kingfisher 30, another bilge-keeler, almost identical to Ridgway's *English Rose IV.*

Sailors, navigators, experts of all sorts lined up to instruct him and help him on his way. When he confessed to them his complete lack of sailing experience, none suggested it was absolute madness for a novice to sail to the Southern Ocean and Cape Horn in a bilge-keeled family cruiser. All of them wilfully shoved common sense aside. Some did express doubts about the boat, but they all helped Blyth, eagerly, toward the edge of an abyss that none of them would have approached themselves.

Although they were preparing their boats at the same time, a quarter of a mile apart on the Hamble River, and saw each other often, Blyth didn't tell his former partner of his plans until shortly before they both left. He was too afraid Ridgway would ask him the question nobody else was asking: "What in God's name do you think you're doing?" He concentrated on his preparations and studiously avoided examining the overwhelming weight of reasons against going.

Chay's wife, Maureen, proved his most ardent champion and ally. Although their daughter was just 10 months old at the time of his departure, Maureen too looked only on the positive side. She helped him plan, pushed him on, and organized and packed all his food supplies, giving him an ample, healthy, and varied diet, including packages of paella, tins of haggis and roasted grouse, and a seemingly limitless supply of his favorite candy, Smarties. Once at sea, he ate better than most of his competitors.

On the day of his departure—June 8, one week after Ridgway had set sail—the 27-year-old Blyth told a *Sunday Times* reporter his reasons for going. "Out here it's all black and white, all survival. I'm not particularly fond of the sea, it's just a question of survival."

Few people leaving a dock for an afternoon's sail in a dinghy have cast off with less experience than Chay Blyth had when he set off to sail alone around the world. Overwhelmed by the details of outfitting his boat, he never managed, as he had once hoped, the OSTAR's 500-mile qualifying passage (not required under Golden Globe rules, which, in the interest of including all-comers, conveniently presumed a certain level of competence). Chay Blyth had sailed no more than 6 miles by himself, and that in mostly calm conditions.

Friends came aboard *Dytiscus III* that Saturday morning, raised and set the sails and the self-steering gear, and then got off, while others went ahead in another sailboat so that Blyth could copy their maneuvers as he left port under the gaze of television cameras. But with the wind vane steering the boat, there was nothing for him to do, so he sailed away with his hands in his pockets until he was out of sight.

Then he discovered he was lost. Out of sight of land for the first time, his hastily acquired navigational techniques deserted him. He knew only that he was somewhere in the English Channel, and that the Atlantic lay to the west. He steered in that direction. Five days later he saw and correctly identified the French Ile de Ouessant at the western end of the Channel. He sailed on into the Atlantic.

Three weeks later, on July 1, having found his way to Madeira and beyond, he sailed into an unseasonable gale and got his first taste of the bilge-keeler's behavior in bad weather. Running before the wind, the self-steering vane would not hold the boat on course. Steering by hand, Blyth couldn't do much better. The two shallow bilge keels lost their grip in the tumbling water near the surface and *Dytiscus III* began broaching uncontrollably: slewing sideways out of control with one wave, to be smashed into by the next. The boat became unmanageable. Nothing Blyth did seemed to help.

> So I lowered the sails . . . and once I had lowered them there was nothing more I could do except pray. So I prayed. And between times I turned to one of my sailing manuals to see what advice it contained for me. It was like being in hell with instructions.

As the two paratroopers got underway, Robin Knox-Johnston was nearing his own departure date. He had hoped to sail on June 1, the earliest date allowed under the Golden Globe's few rules, but the thoroughness and difficulty of his preparations delayed him. He did most of this at Surrey Commercial Docks on the south shore of the Thames River on the outskirts of London. It was a rough but practical location, cheaper than any yacht yard and close to his parents' home in Downe, Kent, where he was living.

One of the most crucial jobs for any single-hander is the installation of self-steering gear. If sailing meant sitting in a boat's cockpit and steering day and night, eyes glued to a reeling compass, with only quick dashes to the galley for food, few people would ever go to sea for pleasure. Even on a fully crewed boat, a two-hour watch at the helm is the most onerous and boring task afloat, and sailors have brought all their ingenuity to inventing and rigging systems to avoid it. Today, most yachts employ battery-powered automatic pilots, but the batteries needed to drive them weren't available in 1968 and the Golden Globe racers couldn't hope to use them without carrying large generators and enormous tanks of fuel. Instead, they employed mechanical, wind-powered self-steering gears. These often look like small weather vanes designed by Rube Goldberg, but when there is any wind at all they manage to keep a boat heading in approximately the right direction. The wind is natural and free, and a device that makes such clever use of it has something of a serendipitous wonder to it. It's not surprising that sailors, particularly single-handers, tend to anthropomorphize their wind vanes, give them nicknames, and talk to them with affection—and sometimes irritation.

They are simple in principle: A wind vane, rotating on a vertical axis like a weathercock, is linked to a small trim tab attached to the aft (or trailing) edge of the rudder, like the flap on an airplane's wing. When the boat is on course, the wind vane is set so that its leading edge points into the wind, offering no resistance. When the boat veers off course, the wind vane presents its side to the wind, which then pushes it around its axis, and this movement is transferred by linkage to the trim-tab, which, like the wing flap, moves the entire rudder, steering the boat back onto its original course. The engineering of such gear can be crude and inexpensive or elegant and costly, but the arrangement is simple, generally robust at sea, and, to the single-handed sailor facing a voyage of tens of thousands of miles, a vital piece of equipment.*

Normally, this gear is attached to a boat's stern, directly over the rudder. On *Suhaili,* however, this couldn't be done, because her mizzen sail and boom stuck out over the rudder and would have interfered with a wind vane there. Knox-Johnston finally designed his own gear, which consisted of two wind vanes, each mounted on steel-tube outriggers on both sides of the boat, the linkage being ropes running through sheaves to the stern. It was at best an awkward arrangment, the outriggers and rope linkage interfering with his movement on deck, but it was entirely in keeping with the boat's rough-hewn character.

Though she was only 2 feet longer than Ridgway's and Blyth's boats, *Suhaili* was twice as heavy as the bilge-keelers. She displaced double the volume of water; she was, literally, twice as much boat. Some of this greater weight came from her massive all-teak construction (*Suhaili* might easily have damaged the trawler that knocked into Ridgway's *English Rose IV*), but most of it was in the form of greater hull volume, into which Knox-Johnston was able to pack an immense amount of food and seagoing stores.

Divorced and having no one to worry about his lonely dinners, Robin Knox-Johnston opted for the standard yachtsman's diet of the day—tins of corned beef or baked beans, grub the avid weekend sailor didn't mind subsisting on from Friday to Sunday nights—and then factored a ghastly 300-day-plus multiplication. Such fare reflected the stolid dreariness and paucity of the English postwar diet, compounded by Knox-Johnston's years of eating the institutionalized food aboard British merchant ships. He loaded *Suhaili* with over 1,500 tins, each one stripped of its paper label, varnished (against rusting by seawa-

*This describes the trim-tab vane gears used at the time. Modern wind vanes often use slightly different principles.

ter), and coded. This was the time-honored practice espoused by English yachtsmen long used to sailing leaky wooden boats.

His staples were:

216 tins of corned beef
144 tins of stewing steak
48 tins of pork sausages
72 tins each of green beans, runner beans, carrots, and mixed vegetables
144 tins of Heinz baked beans
48 tins of Heinz spaghetti in tomato sauce
216 tins of condensed milk
40 tins of processed cheese

And tins of fruit, jam, salad dressing, cooking fat, soup, and much, much more.

To cram it all in, Knox-Johnston tore out the bunks in *Suhaili*'s forwardmost compartment, the fo'c'sle, and built shelves and lockers. More food was packed in 5-gallon jugs and containers, until there were jugs, drums, crates of food and drink filling the floorboard space between the main cabin bunks, jammed into the cockpit, and packed into every conceivable space in the boat. He also took aboard a small ship chandler's warehouse of stores, including tools, extra sails, lines, rigging wire, anchors, jugs of kerosene, diesel, and gasoline, and spare parts for every device aboard. The boat was prepared for an almost indefinite stay at sea.

Despite being a bookworm, Knox-Johnston's reading of the classics had been spotty. He now had the ultimate opportunity, that unlikely "someday" suddenly looming before him. Dr. Ronald Hope of the Seafarers' Education Service provided him with a boatload of such works as *Tristram Shandy, Vanity Fair, War and Peace, Crime and Punishment, Tom Jones, Clarissa, History of Western Philosophy,* and instructive works like *Chess in a Nutshell, A Textbook of Economics, and Elementary Calculus.* There was much more, but like his diet, it was unrelievedly solid stuff.

He also brought along a correspondence course for the Institute of Transport examinations. This was a conscious decision to give his mind additional exercise. In the same spirit, his sponsor, the *Sunday Mirror,* sent him to a psychiatrist so that his state of mind could be compared before and after his voyage. The psychiatrist declared him "distressingly normal."

His "normalcy" was no doubt a distress to the psychiatrist, but the diagnosis was fundamentally mistaken. Normal people aren't driven to try to sail alone around the world without stopping. They don't stop their lives midstream and embrace, with single-minded effort and every resource available to them, a hair-raising stunt never before attempted and which has every chance of killing them. None of the Golden Globe sailors could have been called normal, but alone among his rival competitors, there seemed to be no dark streak of introspection in Knox-Johnston's forthright personality. He was what might pass for normal in some sunny world where human nature is not best defined by its aberrations from a hypothetical model.

On June 3 he sailed from Surrey Commercial Docks. Waving good-bye to him from shore was his 5-year-old daughter Sara, the product of his marriage that had crumbled in India under the strain of a seaman's absentee lifestyle. He had gotten to know her well since returning to England, and they had spent his last weekend ashore together. It was a wrenching moment for him, and he hoped it wouldn't be too bad for her. But she was used to him departing for long periods, and he was relieved when she stopped waving and began playing with a small radio.

Carried on the tide down the sea reach of the Thames, *Suhaili* presented a curious sight: with her two wind vanes (which Knox-Johnston had painted Day-Glo orange for visibility) mounted on steel outriggers on either side of the hull, she now had the appearance of some strange homemade fishing boat. And loaded with a year's supply of food and water, she floated low in the water and was sluggish underway, her already slow hull-shape slower. Knox-Johnston could only hope she would pick up speed and buoyancy as he neared the Southern Ocean and ate his way through his cargo.

He took with him a crew of three, including Bruce Maxwell, a reporter from the *Sunday Mirror,* and a *Mirror* photographer, to give him a hand on one of the most perilous sections of his entire voyage: the English Channel, which is so congested with shipping that traffic through the Strait of Dover is monitored on radar screens by controllers just as it is at airports. *Suhaili* was bound for her final port of departure, Falmouth, Cornwall, the most westerly harbor of any size in southern England, a traditional port of departure in the age of sail because of its easy access to the open Atlantic. A pretty town tucked inside the wide mouth of the river Fal, protected from all weather by the sheltering arms of green hills, Falmouth has a long association with ships and the sea. It's a good place to do last-minute work on a boat and an auspicious port of departure.

Knox-Johnston and his crew reached Falmouth on June 9. There they spent 5 more chaotic days in port attending to a mass of final details.

He sailed on June 14 (a week after Chay Blyth), giving further evidence of how far his normalcy deviated from center: it was a Friday. This was a remarkable flouting of a seaman's superstition. Luck is an acknowledged but unquantifiable factor at sea; Knox-Johnston clearly believed in making his own. He was ready, the weather was fair, his boat was slow, and he didn't want to waste a single day, so he sailed.

"Rifle-and-Bible Seaman Sails" reported the *Sunday Times,* noting that just before he left Falmouth, his churchwarden father handed him 100 rounds of .303 ammunition for the rifle he carried aboard, and the port chaplain had taken him into town to replace the bible Knox-Johnston had left at home. *Suhaili* was still "a bit of a bog inside," Knox-Johnston told the reporter, who wrote that the boat's crude outrigger wind vanes gave it a "wallowing, trawler appearance." But a canny harbor official, wittingly or otherwise perceiving *Suhaili*'s Nordic antecedents, observed that she was "a real old ice-breaking boat. If she hit England, I'd be concerned for England."

Launches carrying the press and his family followed *Suhaili* for a few miles beyond Falmouth before turning back. The wind was light and from the northeast, and the boat crept slowly southwest away from land. During Knox-Johnston's first night at sea, as he sailed among the shipping funneling into the English Channel, he dozed in the cockpit clutching a flare, ready to alter course and alert ships to his presence.

His first week at sea proved disappointing. The wind remained light and the heavily burdened *Suhaili* made daily runs of 77, 80, 52, 38, 62, 87, and 100 miles. But Ridgway and Blyth, far less experienced sailors than Knox-Johnston, had the same weather and their mileages were even less. Though none of the three knew it at the time, Knox-Johnston began to gain on them.

On Sunday, June 30, the *Sunday Times* carried a full page covering its Golden Globe race. It announced a new competitor: Royal Navy Lieutenant Commander Nigel Tetley, 45, also unsponsored, who planned to sail in September aboard the trimaran that he and his wife used as their home.

The article briefly described all the competitors and their boats. And it quoted Sir Francis Chichester's hope that every entrant in the race should have a proper briefing on equipment, provisions, and the conditions that would likely be encountered: "We can't stop anyone who is determined to set out, but we can make sure he knows what the game is all about."

This might have been a good idea, but neither Sir Francis, nor any representative of the Golden Globe race briefed or advised the men who set out. The *Sunday Times* simply sent its reporters to cover what the sailors said or did, and camera boats to jockey for position among its competitors. Yet the newspaper didn't hesitate to admonish those whom it felt showed a want of instruction.

> Three sailors have already left England ill-advisedly—most seasoned yachtsmen agree a June start is almost bound to spell ugly weather in the Southern Ocean. . . . All three [Ridgway, Blyth, Knox-Johnston] are young men.
>
> The canniest sailors—Howell, Commander Bill King, the former submarine ace, Loïck Fougeron and Bernard Moitessier, the dry, calm French single-handers—will all leave during July and August to ensure the best possible weather conditions. . . . Commander Nigel Tetley . . . will be [an] end-of-the-season starter. And there is one mystery competitor—Donald Crowhurst, a 35-year-old manufacturer of navigational equipment, who will reveal nothing about his new ketch for fear that other yachtsmen might copy some of its "revolutionary" ideas. He . . . and Tetley are [both] experienced sailors; it seems that the younger, rasher men have started first.

S

6

ONLY THREE MONTHS EARLIER, a fifth competitor (counting Bernard Moitessier, preparing in Toulon) had decided to join the race.

Sunday, March 17, 1968, was cold and wintry along England's south coast. Nigel Tetley and his wife, Eve, read the Sunday papers in bed at home—a 40-foot trimaran moored in Plymouth Harbour. Living full-time aboard a trimaran was not the hardship it could be on other yachts: *Victress* was 22 feet wide and had the living space of a small cottage. That morning they had a small coal stove burning inside the cabin, which was toasty and comfortable. Tetley, 44, was a commander in the Royal Navy, less than a year from retirement, and Eve, his second wife, taught geography in a local school in Plymouth.

While Eve looked through one paper, Tetley noticed a headline in another: "Round-the-World Race." He picked up the *Sunday Times* and read the details of the Golden Globe race.

Tetley had once thought of sailing alone around the world, but two years earlier he had met Eve and, like many a would-be single-hander, he'd given up ideas for solo adventures.

But he surprised himself. The notion of this race tapped into a compulsion stronger than he was aware of. He knew immediately that he wanted to do it.

He pushed the paper across the bunk so that Eve could read it. When he was sure she had, he said, "May I go?"

After a long look at her husband, Eve said, "I would not try to stop you."

Time was now short to launch such an effort and Tetley began to prepare immediately. Like Commander Bill King and Robin Knox-Johnston, he believed his best chance lay in having a boat designed and built specifically for this race. *Victress* was not a racing boat but a five-year-old live-aboard cruising yacht. Tetley had sailed her, with crew, in the 1966 Round Britain race, and the trimaran had performed well, finishing in fifth place. But a few weeks going around the British Isles in summer was small beer compared to four or five months in the Southern Ocean. What he had in mind now was an altogether different machine, a fleet, simple, powerful 50-foot trimaran, that would cost him £10,000, money he didn't have.

He began writing letters to possible sponsors. He started with Tetley Tea, hoping they liked his name and reminding them of the long association between tea and the clipper ship route of the race. The tea firm declined. He wrote to other tea, tobacco, and drink companies but the result was the same. He had nothing to offer a sponsor. He was not, like Ridgway and Blyth, a newly famous adventurer. He was certainly a better seaman than either of them, but an unknown if competent yachtsman was not what an advertiser was looking for.

Nor could he get away quickly. Tetley was still on active naval duty ashore in Plymouth, though he believed the navy would release him in September, five months before his official retirement date. This would enable him to leave before the *Sunday Times* deadline of October 31, but he would be starting later than the rest of the fleet. His only chance of winning would be sailing faster than any of the other boats, possible in a trimaran.

Late in March, waiting for replies, the Tetleys set out on a short cruise aboard *Victress* with Nigel's two sons from his first marriage. While they were trying to berth the boat in windy conditions in Penzance Harbour, Cornwall, a section of the trimaran's port bow was damaged. All the local shipwrights were busy fitting out boats for Easter. The only person Tetley could find to repair the damage was a coffin maker.

Sailors are naturally a superstitious lot. When they head out upon the deep, the constructs of society soon drop astern and they are surrounded by shooting stars overhead, phosphorescence in their wakes, and heaving shapes all around them in the sea and sky. It is easy, then, sensible even, to become afraid. Sailors are generally careful and conscientious engineers, fussing endlessly with their craft to make them seaworthy, but when they have done all they can do, they move quickly and easily to prayer. They have many of their own, like Psalm 107, surely written by a sailor.

They that go down to the sea in ships, that do
business in great waters;
These see the works of the Lord, and his wonders
in the deep.
For he commandeth, and raiseth the stormy wind,
which lifteth up the waves thereof.
They mount up to the heaven, they go down again
to the depths: their soul is melted
because of trouble.
They reel to and fro, and stagger like a drunken
man, and are at their wit's end.
Then they cry unto the Lord in their trouble, and
he bringeth them out of their distresses.
He maketh the storm a calm, so that the waves
thereof are still.
Then are they glad because of the quiet; so he
bringeth them unto their desired haven.

Prayers may comfort. They may even work. But, as backup, in the spirit of assisting the Lord in his aid of those who help themselves, sailors carry with them a seaman's chest of superstitions, evil eyes, rites, and invocations handed down from Homer's time. Some are specific: never sail on a Friday; never whistle aboard a boat unless it is for wind when becalmed; always make an offering to Neptune on crossing the Equator. One precaution should have appeared obvious, not only to the mariner: never have your ship repaired by a coffin maker. But Tetley wasn't perturbed, so the coffin maker did the job.

Facing a mounting pile of rejection letters from hoped-for sponsors, Tetley soon decided, like Knox-Johnston, that if he was to go at all, it would be in the boat at hand, *Victress,* newly repaired. Still, he needed money to refit his boat for the voyage.

It wasn't until the third week in June, after the first three "rash" young men had departed, that he announced his intention to sail, without sponsorship, to the *Sunday Times.* A reporter and a photographer came down to Plymouth to interview and photograph the latest entrant. Tetley mentioned to the journalists that he and Eve enjoyed listening to classical music while sailing, and he demonstrated the trimaran's stereo tape system. The reporter suggested he look for a record company to sponsor him.

The following weekend, on June 23, the *Sunday Times* carried an article with the headline, "Around the World in 80 Symphonies," with a photograph of Tetley and his wife Eve, both attractive people, laughing at *Victress*'s saloon table. The article noted that the Tetleys shared a love of music, and that the commander, hoping for a music company sponsor, "wouldn't even mind being plugged as the 'Around the World in 80 Symphonies' mariner."

Richard Baldwyn, director of Music for Pleasure, a company that marketed cassette tapes, was reading this article while flying back to England from the south of France. Approaching England, the pilot announced that there was a problem with the plane's landing gear: the wheels might not go down. Baldwyn offered up a prayer to the effect that if he reached the ground intact, he would sponsor the music-loving Tetley. The plane landed safely, and Tetley got his sponsor.

He and Eve worked every spare hour together preparing *Victress* at Plymouth's Millbay Docks. This meant afternoons and long into most nights since both were working during the day—Tetley still had naval duties, and would not be released from service until September 1.

Eve organized Tetley's food with great attention to nutrition, variety, and appeal. She augmented the inevitable corned beef and staples of the sailor's diet with tins of roast duck, roast goose, jugged hare, smoked turkey, venison, roast pheasant, and a huge variety of seafood, including octopus and rainbow trout.

Plymouth is a large natural harbor, the home of the British Navy, an ideal place in which to prepare any vessel for sea. Tetley was soon joined by three other Golden Globe racers. The first arrived one day in July. Nigel Tetley looked up as the lock gates at Millbay Docks opened and a rough-looking steel ketch sailed in. It was unlike any yacht he had ever seen. There was a brutal practicality to it: telephone poles for masts, steel pipe for its bowsprit, no wood to varnish, just a plain paint finish—the only hint of warmth was the startling fire-engine red of its hull. A wiry, muscular figure stood at the bow rolling a cigarette as the boat appeared to sail itself slowly toward the dock. Tetley hailed the newcomer, asking the name of the vessel.

"*Joshua,*" came the reply. Bernard Moitessier had arrived.

In the weeks that followed, two more boats, Loïck Fougeron's steel cutter, *Captain Browne,* and Bill King's brand new light-displacement, slippery-smooth, cold-molded, junk-rigged schooner *Galway Blazer II,* joined *Victress* and *Joshua* at the dock.

Fougeron, a 42-year-old Breton who managed a motorcycle company in Casablanca, was a friend of Bernard Moitessier's. He had sailed with him aboard *Joshua* from the Moroccan coast to the Canary Islands and undoubtedly was heavily influenced by the French sailing superstar. Fougeron had an able vessel, a steel-hulled 30-footer that would have met with Moitessier's approval.

Before sailing *Captain Browne* from Toulon, where he had fitted it out near Moitessier's *Joshua,* to Plymouth, Fougeron had had almost no single-handed experience.

The four sailors quickly put aside the concerns of rivalry. They looked over one another's boats, freely traded information, ate dinners together, and talked about their race. They established the camaraderie of soldiers waiting to ship out for war.

S

7

IN THE MIDDLE OF JUNE, two weeks out of Inishmore, John Ridgway was closing with Madeira where he had arranged to meet a journalist, Bill Gardner, from his sponsor *The People,* and hand over letters and photographs. He was making good time, but his mood was all wrong. The loneliness most single-handers quickly adjust to had only intensified. His hand-cranked Lifeline radio, the same model he had used successfully on the transatlantic row, had stopped working; unable to send or receive messages to and from home, his loneliness grew even worse. A compulsive eater ashore, he was losing his appetite and had to force himself to eat his daily "rations" prepared for him by the Horlicks company. And the collisions at the start of his voyage had tapped into a deep wellspring of anxiety. He confided to his log that he was now hearing, above the constant groans of a boat underway, "ominous creaking sounds" from the area at the side of the boat that had been hit by the trawler. Ridgway's confidence in himself and his boat was ebbing away.

On Sunday, June 16 (two days after Robin Knox-Johnston sailed from Falmouth), Madeira's mountainous profile rose out of the sea ahead. Fishermen in small boats saw Ridgway and waved. The land turned green and terraced with fields as he approached. But in the afternoon as he neared the northwest corner of the island, the spot for his rendezvous with Bill Gardner, a strong local wind, skewed and reinforced by its passage around the high land, rose and blew him offshore, where he reduced sail and hove to for the night.

The next day, a local boat carrying Gardner found Ridgway and *English Rose*. The men waved and shouted at each other across the water through loud-hailers, and then Gardner's boat drew close enough for normal talk, even jokes.

"I've been waiting ten days," said Gardner.

"I'll bet you have. Sunning yourself on the beach."

For a precious few minutes Ridgway had the companionship he sorely missed.

Using waterproof canisters pulled through the water by line, he sent Gardner a package of diaries, films, and tape recordings on a line. Gardner sent him back letters, newspapers, and local Madeiran bread, cheese, sardines, and beer. They talked about the race, Gardner filling him in on when Chay Blyth and Robin Knox-Johnston had sailed.

Soon it was time to part. Ridgway asked him to send his love to Marie Christine and his baby daughter, Rebecca, and the men agreed they would see each other again off the town of Bluff, New Zealand, in October, three and a half months away. Then Gardner's boat motored away.

Ridgway headed for the open sea, enveloped once more in absolute isolation.

Later, reading a copy of the *Sunday Times* that Gardner had passed to him, Ridgway discovered that the race rules governing "taking on supplies" extended to the mail and the fresh lunch he'd just received. How ridiculous! he thought, enraged by the pettiness of it. Obviously Gardner, who had now technically disqualified Ridgway, had thought so too.

Ridgway plugged on, but he remained desperately lonely and unhappy. After a struggle getting a sail down in windy conditions one morning, he came below into the cabin and burst into tears.

He realized that he had cried at some point on each of the last 27 days.

He wondered why he was attempting this voyage, and what had driven the other competitors to attempt it. Years before, when he had thought of giving up the canoe race after capsizing, it had been his partner Chay Blyth who had been fierce about not giving up, who had pushed them on and made them win. On their transatlantic row, it was Blyth who kept up their spirits, once repeating over and over during a 5-day storm, "It's almost over, soon it'll be a memory."

Ridgway later wrote:

> Whenever we were really miserable Chay would strike up with the old Scottish songs of his childhood. "The Road and the Miles to Dundee" never failed to rally my spirits; he was tremendous when things seemed really grim.

When they were closing with the coast of Ireland and worried that a storm might throw them against the cliffs, Ridgway had offered to make a call for help on their radio. "We'll go on," said Blyth unhesitatingly.

Now Ridgway suspected that on his own he wasn't hungry enough to win.

A seaman is not made by simply going to sea. He must also find in himself a love for it. Ridgway was not engaged by the sea. He had no feeling for it, no love of its literature, no sea heroes to emulate. As with his transatlantic row, the sea was simply a hostile environment to be survived, the voyage an ordeal to be endured. Ridgway kept his mind ashore. He thought of home. He thought of the adventure school in Scotland that he hoped to start with Marie Christine. He listened to broadcasts of cricket test matches on the BBC World Service and vividly remembered his own visits to Lord's cricket ground in London, and "the great bags of cherries eaten very slowly in the stands, while white figures dashed about the green grass, far below. The pigeons, muted applause, the scoreboard—I could see it all."

He pushed *English Rose* south, but his heart was no longer in the voyage.

A few hundred miles astern of Ridgway, Chay Blyth was in much the same frame of mind. He was also having trouble getting his radio to transmit, his boat was worrying him, and he was lonely. He had a further problem: he was still uncertain of his position. His early efforts at celestial navigation had him on dry land smack in the middle of an island among the mountainous Cape Verde group (6,000 feet high), yet he could see nothing but empty ocean all around him.

He wondered how Ridgway was doing, enviously supposing that his former shipmate was untroubled by the same worries and loneliness. But Blyth had only seen Ridgway in the company of Blyth, and he did not imagine that his superior officer might be going to pieces without him, without the obdurate driving force of Blyth's personality.

Yet the image of Ridgway doing so much better was a boon to him—as were all his difficulties. Adversity was like an electric cattle prod to Chay Blyth. It spurred him on. His reaction to difficult or desperate conditions had always been the opposite of Ridgway's.

Ridgway, a deeply introspective man, was acutely aware of his tendencies toward weakness and softness. He had fought this by boxing at school and in the army and by driving himself to become as hard and tough as he could be. He believed it was his own strength or weakness that would bring him success or failure. He counted only on himself. Chay Blyth had a far simpler view. His efforts would count for only so much; after that it was up to God. And God was on his side, he had no doubt of that. When he prayed during bad weather and the weather subsequently got better, he knew why: "It has died down a bit now," he wrote in his logbook, "after I had prayed for it to go down. Nobody on this earth could convince me that there is no Lord."

Now, in the middle of the Atlantic, lost, lonely, and bruised inside his lurching weekender's cruiser, Chay Blyth was unavoidably becoming aware of his boat's shortcomings, its inherent unsuitability for the Southern Ocean. He was beginning to realize that at some point ahead, probably around the time he entered the Southern Ocean, he would face a choice between persisting with the voyage despite being both fundamentally unprepared and in the wrong boat or giving up. Yet for the time being he pushed on with savage determination. He embraced almost exuberantly everything that was thrown at him with all the toughness his years of army training had bred in him.

Blyth was getting a feel for the sea. He was becoming a seaman.

In contrast, Robin Knox-Johnston's contentment at sea was striking. He had already spent two years living aboard *Suhaili,* in port and at sea, and he felt completely at home in her. He quickly and comfortably fell into a seagoing routine adapted from his 10,000-mile voyage from India.

Unlike most sailors, solo or sailing with others, Knox-Johnston enjoyed jumping overboard for a swim if the day was warm. The perimeter of one's vessel at sea is the very clear boundary of safety; beyond it lie peril and possibly death, and their nearness is keenly felt. It helps keep one aboard. The visceral, unreasoning fears of abandonment, the abyssal depths and all its creatures, make it very difficult for most people to jump overboard once the shore has receded beyond a short distance away, even with trusted companions aboard a boat. On one occasion when Ridgway and Blyth were rowing across the Atlantic, one of them had to go in to see if the boat's rudder had been damaged. The weather was hot, the sea was calm, yet they argued all day to see who would go overboard. Eventually Ridgway dove in, inspected the rudder, and got out fast. "Go on, Chay, it's lovely," he said, grinning. Blyth went in and was quickly back on board. Neither of them went in again.

Knox-Johnston was untroubled. Trailing a line astern, he

would jump from the bowsprit and swim alongside *Suhaili* until she overtook him, then grab the line and pull himself aboard. This, he felt, kept him fit and clean.

He might have been on vacation.

> A sedate lunch followed my swim, usually consisting of biscuits and cheese or the like, with a pickled onion on special occasions as a treat. The afternoons would be spent just like the morning, working or reading, until 5 P.M. when, if I felt like it, I dropped everything for a beer or a whiskey.

And from his logbook:

> I repaired the Gilbert and Sullivan tape cassette . . . and had a wonderful evening. I joined in sitting at the table in the homely light of the cabin light. It is not cold enough yet for clothes, just pleasant. . . . I think I'll have a nip of Grant's. I can think of no one with whom I'd trade my lot at present.

He was by now a seasoned navigator, with a seaman's knowledge of the ocean's wind and current systems, and he made steady if unremarkable progress south, gaining on the two army sailors.

Like all solo sailors, he had to determine how long he could sleep before waking to check that he was not about to be run down by a ship. Figuring the likelihood of encountering a ship at sea is impossible. Until the middle of the twentieth century, most ships kept to well-defined shipping lanes, routes across oceans that offered the most favorable combination of weather and ocean conditions, and economical distance run. The British Admiralty publication *Ocean Passages of the World,* which the Golden Globe racers carried aboard their boats, is a pilot book detailing these preferred routes for high- and low-powered vessels, as well as for sailing vessels. It comes with charts showing these shipping lanes.

Early single-handers—and the sleepy shorthanded crews of other small sailboats—could avoid these highways in the sea or, if they had to cross or approach shipping lanes, knew what to expect and could remain awake or catnap for a few days. Afterward, out of harm's way, they could, and usually did, turn in for hours at a time. But about the time of the Golden Globe race, ships began to stray out of their lanes. They became more powerful, able to head more directly for their destinations against prevailing winds and currents. They began to get daily radio reports from shore stations giving them optimum courses around local weather systems.

At one time, the single-hander under sail could hope that the approaching ship would see him and alter course, as it is legally obliged to do: a vessel under sail has right of way over an engine-driven ship. Sailors could reasonably expect that any oncoming vessel would have a man in the bow peering out into the dark ahead who would see their little light and send a message back to the bridge, and the ship would turn away. But by the 1960s, this was increasingly not the case. As ships have grown larger and their systems more sophisticated, manpower aboard has been cut back. A supertanker may have fewer than twenty men aboard, and at any given time a third of that complement will be off duty, asleep, or below reading. A few shipping lines still maintain a good lookout, posting a man on the bow in radio contact with the bridge. Other ships, particularly those registered under the less demanding requirements of flags of convenience, are not so scrupulous. Lookout may be by radar alone, and if the radar doesn't pick up a boat, it's invisible. Yachts, particularly wooden yachts, do not make good radar pictures. They're small, their radar echoes may be lost in "sea clutter"—just more waves on the radar screen. And as sailors often find when calling a ship by radio to ask what sort of radar picture their boats make, the radar may be turned off.

The bridge of a large tanker may be a quarter of a mile astern of its bow and 150 feet above the water—something like the view from the upper floors of a condo in Miami Beach looking out at the Florida Straits. The crew on the bridge can see the big stuff, other ships, from up there, but little sailboats can go unnoticed. At night, a sailboat's navigation lights, close down to the water, will almost certainly not be seen farther than half a mile away, even if anyone's looking—scant minutes to collision. Then, if seen, the maneuverability of a large ship is poor and slow.

The curve of the earth, it soon becomes apparent at sea, is quite pronounced. The horizon seen from the deck of a small yacht is about three miles away. Beyond 3 miles, a ship will be "hull-down" below the horizon: only its superstructure is visible. Eight miles away, the whole ship will be below the horizon. Conditions of haze, cloud, rain, fog, or a large swell on a sunny day can reduce this to yards. A ship moving at 18 knots (the speed at which the average container ship might travel; many travel faster), unseen when the sailor comes on deck to make a careful scan of sea before going below again, can steam up over the horizon and run a yacht down in 20 minutes or less.

Clearly, most sensibly, it's up to the sailboat to stay clear of the ship. The single-hander, therefore, must wake, climb on deck and look around every 15, 20, 30 minutes—there is no rule, it varies from single-hander to single-hander.

Knox-Johnston, the merchant seaman who had been trained aboard rigorously well-run British ships, had a touching, old-fashioned faith in the idea that all ships maintained a lookout. Before his voyage was over, this faith would be shattered. Near land and shipping lanes, he dozed in the cockpit, ready to wake and alter course. Midocean, he was as untroubled by doubts as he was when swimming, and tended to sleep for hours at a time when the weather was fine and *Suhaili* didn't need attention.

But *Suhaili* was not without her problems. Even in quiet weather, the bilges were filling with water, and Knox-Johnston was having to pump them dry twice a day. She had leaked before, on the voyage from India, and he had noticed it again on the run from London to Falmouth. Now the leak was worse. A little water in the bilges is not uncommon for any boat, and more the rule for wooden boats of conventional plank-on-frame construction like *Suhaili*. But the amount of water now flowing into the boat was significant, and Knox-Johnston was worried that this might indicate a weakness in the hull.

Becalmed south of the Cape Verde Islands, he pulled on a mask and snorkel, jumped overboard, and swam down underwater to inspect the hull. The trouble was immediately apparent: a long gap showed in a seam between planks just above the keel, near the spot where the foot of the mainmast was anchored into the keelson. There was a similar gap in the same place on both sides of the hull. As *Suhaili* rolled slightly, Knox-Johnston could see the seam opening and closing with each roll. He surfaced, hauled himself aboard, lit a cigarette, and considered the problem. He was worried that the floor timbers—not the floorboards, but the thick-sawn members that joined the hull frames to the keel and held the bolts that kept the heavy iron ballast attached to the bottom of the boat—might be weakening. A failure with the floors could be catastrophic, even resulting in the bottom of the boat coming apart and falling off. Most of the floors were covered by water tanks built into the boat, but Knox-Johnston poked around in the bilges, inspected those he could see, and did what he could to convince himself that the floors were not failing. It was simply a caulking problem, he decided—the only problem he could realistically fix.

Having convinced himself that caulking—hammering twisted lengths of cotton into the seams and covering it with (normally) putty—was the answer, he had to figure out a way of doing it 5 feet underwater. He tied a hammer on a line and lowered it over the side to dangle in the right place. Then, dressed in a dark shirt and jeans to hide the whiteness of his body from any cruising sharks, he went overboard.

The job was impossible. He tried hammering the cotton into the seams, using a screwdriver in place of the traditional caulking iron, but it wouldn't stay in the seam and came out every time he surfaced for air. After a fruitless half hour he climbed back aboard.

He decided to sew the bead of cotton onto a long, narrow strip of canvas, which he then coated with Stockholm tar to stiffen it. Then he pushed copper tacks through the canvas. He lowered himself over the side, swam down underwater, held the long strip with the bead of cotton inside the seam, and started hammering the tacks into the planks. After repeatedly rising for air and diving back down, the long band-aid was finally nailed in place. But for how long? He worried that the canvas would wear away. It needed a tougher covering. Back aboard, he made a strip of copper from long sheets left aboard by the Marconi engineers who had installed his radio. On deck, he hammered tacks through the copper, intending to go overboard again and nail it over the canvas.

First, to warm himself after two and a half hours in the water, he made a cup of coffee. While drinking it in the sun, he noticed a dark gray shape cruise close by the boat: a shark. He watched it, hoping it would go away: killing it could attract more sharks. But ten minutes later it was still circling the boat, so he got out his .303 rifle, threw some toilet paper into the water, and waited. On its first pass, the shark swam below the paper, but then it turned, came back, rising toward the paper. As its head broke the surface, Knox-Johnston fired. The shark convulsed furiously for half a minute and then was still and slid away down into the deep. Two pilot fish that had been swimming with the shark peeled away as it disappeared into the depths and took up station beneath the shadow of *Suhaili*. For half an hour he kept a lookout for more sharks, but then a light wind arose and forced him back into the water to finish his repair. He spent another hour and a half nailing the copper strip over the caulking on the port side, every minute of it expecting a dark shape to appear. By then the wind had risen to agitate the water and push *Suhaili* on, forcing him to leave the starboard side until the next calm. He had been in the water for four hours.

Two days later, *Suhaili* was becalmed again, and Knox-Johnston repeated his underwater caulking job on the starboard side. The leaking stopped almost completely.

8

A YEAR AFTER he had loaned Donald Crowhurst £1,000, Stanley Best, the Taunton businessman who had made his fortune selling caravans, had become disenchanted with the poor sales of the Navicator and the prospects for Electron Utilisation, and he wanted his money back.

In a letter dated May 20, 1968, Crowhurst wrote to him arguing that, contrary to what Best might think now, the company was about to capitalize enormously on his own entry in the Golden Globe race. He was planning to have a trimaran built, and he stood every chance of winning. The trimaran, he wrote, was a new and controversial type of sailboat, poised to become "the caravan of the sea." Moreover, he continued, "the trimaran is a highly suitable platform for the electronic process control equipment. The only equipment available so far is crude and works along entirely the wrong lines. . . . If the practical utility of the equipment I propose can be demonstrated in such a spectacular way as in winning the *Sunday Times* Golden Globe and/or the £5,000 prize and it is properly protected by patents, the rapid and profitable development of this company cannot be in any doubt."

What he said made sense. Multihulls were the coming thing. In 1960 an American, Arthur Piver, had built his own 30-foot trimaran, *Nimble,* out of cheap plywood for $2,000 and sailed it across the Atlantic from Fall River, Massachussetts, to England, with a stop in the Azores, in 28 days at sea at an average of 136 miles per day. This was within hours of Eric Tabarly's (nonstop) record-breaking passage of 27 days, four years later in the 1964 OSTAR. In 1961, Piver sailed from Los Angeles to Honolulu in his 35-footer, *Lodestar,* in 15 days at an average of 150 miles per day. These were unheard-of speeds for cruising boats in the open ocean. In England, James Wharram started building slender-hulled catamarans based on Polynesian craft and sailed them across the Atlantic. Yet these two vanguard designers were considered cranks by the yachting community, which generally discounted the long-term seagoing possibilities of such light, unballasted craft. But then two catamarans did well in the 1964 OSTAR, and the impressive win by Derek Kelsall in his trimaran *Toria* in the rugged 1966 Round Britain race began to swing the balance of opinion. Eric Tabarly took one ride aboard Kelsall's trimaran and decided to build his own for the 1968 OSTAR.

Multihulls had too many advantages to ignore: they were more for less. Being far lighter than monohulls, they were cheaper to build. Large and spacious inside, with double bunks and large galleys, they sailed upright with very little heeling, and they gave their crews a drier, more comfortable, less frightening ride than traditional sailboats. Many sailors whose wives had not enjoyed sailing found that they were happier and more willing to come for a weekend cruise aboard a catamaran or trimaran. And multihulls sailed fast—two or even three times the speeds of single-hulled vessels of the same length. They were the racing vessels of the future.

Multihulls had one downside, however: without the ballast keel and self-righting properties of traditional sailboats, once they flipped over they stayed that way. If anything, they were more stable upside down than right side up, with their mast and sails poking down into the depths acting as a splendid lightweight keel. This only happened rarely, when an overcanvased boat was pushed too hard by a racing crew. But a primary piece of equipment that Donald Crowhurst was working on, he told Stanley Best, and would employ in his vessel, was a revolutionary, electronically activated, self-righting mechanism that would prevent such a capsize.

If Stanley Best had asked anybody about trimarans in 1968, they would have confirmed most of Crowhurst's claims. A new generation of multihull designers and builders was appearing all over England, and their boats and blue-water voyages were making news. The English Prout brothers were building some of the largest catamarans afloat, and their designs were among the most popular in Europe. Tabarly had sparked the interest of the French, who are perhaps quicker to embrace change and less burdened by traditional boat aesthetics than the English. They were soon enthusiastically sailing trimarans and catamarans around Europe and across the Atlantic. Multihulls were poised for growth, and, if Crowhurst's gadgets worked well on a fast sail around the world, Electron Utilisation could hop aboard for the ride.

But Crowhurst had never sailed aboard a trimaran. His knowledge of them came from magazine articles. Seeing a mounting field of other Golden Globe competitors in the final stages of preparation, his sudden enthusiasm for trimarans was born of urgent necessity. If he was to have a boat built, it would have to be cheap. And he would depart late, at the very back of the field, so it would have to be very fast. A trimaran was the logical solution.

At the moment when Stanley Best was about to sever his connection with him, Donald Crowhurst hooked him with his grand idea and reeled him all the way in. Best could not later explain it. "I, who have always invested in a certainty or a rigorously calculated risk, suddenly jumped into this mammoth undertaking, which I didn't really comprehend, with only the shadowiest prospect of a proper reward. It was, I suppose, the glamor of the idea, the publicity and the excitement—and the persuasiveness of Donald. He was, when all is said and done, the most impressive and convincing of men."

Sometime in late May or early June, Best agreed to pay the building costs of a trimaran, although he expected Crowhurst to continue to seek other sponsors to share the expenses of the voyage. If Crowhurst won, Best's investment would prove a good one. If he did not win, they hoped that his participation in the race would still prove a highly visible advertisement for the products developed by Electron Utilisation. But Stanley Best attempted to protect himself from total loss: the agreement they drew up stipulated that if Crowhurst did not complete the voyage for some reason, he would have to pay Best back the cost of the boat. This meant that if he failed to make at least a good showing in the race, Crowhurst would lose his business and find himself bankrupt.

This didn't worry him. He was deeply convinced of his ability to pull it all off. This was the great challenge, the supreme showcase for his talents, something he had sought all his life. This would change the long run of bad luck that had dogged his parents' lives and his own. Any outcome less than his sweep of the prizes and the fame and glory was an impossibility he was unable to countenance. If Stanley Best wanted to consider it for the security of his investment, Crowhurst was happy to agree.

He was already negotiating with boatyards. Time was now extremely short; no yard would undertake to build the finished boat from scratch in time for Crowhurst to get away before the *Sunday Times'* October 31 deadline, five months away. But Cox Marine in Essex, a sizable boatyard with a production line of trimaran orders that could not be interrupted, suggested that they could build the three hulls, and another builder, Eastwoods, not far away in Norfolk, could assemble the hulls and complete the boat. John Eastwood and his partner John Elliot agreed. It would be a big job for their small yard, but they recognized and embraced the opportunity. As with the boatbuilders who supplied Ridgway and Blyth with their unsuitable craft, the notion of a nonstop circumnavigation was exciting; any boat that came back from such a voyage would bring fame and business to its builder.

The boat Crowhurst wanted was a *Victress*-class trimaran, 41 feet long overall, 22 feet in beam—a sister ship to Nigel Tetley's trimaran. Cox Marine was already building another *Victress,* so they were ideally set up to pop out another three hulls at short notice. The *Victress* was another design by the pioneering American Arthur Piver, whose reputation hadn't been hurt too badly by his own disappearance and presumed death at sea aboard one of his own boats in 1968. His boat might have capsized, but it might also have been run down or suffered any of the mishaps sailors always risk when going to sea.

While Crowhurst and his building teams were working out details, they read in the June 23 issue of the *Sunday Times* that Royal Navy Commander Nigel Tetley intended to enter his *Victress* in the Golden Globe race. Tetley's departure date was given as sometime after August. This quickened all their pulses, but Crowhurst was not perturbed. His revolutionary self-righting equipment would mean that he could push his boat harder, closer to the edge, than Tetley would dare and, if it happened, recover from a capsize and sail on.

In a very short time, Crowhurst's dream had started to come true. He had found the money and construction had begun on his boat. Such was the power of his ideas. The hardest part, the convincing of others, was now behind him. What remained was simply the implementation the plan, and Crowhurst had drawn up a chart that "proved," mathematically, the near-certainty of his double win.

Entrant	*Likely Max. Speed*	*Highest Probable Average Speed*	*Departure Date (P=probable)*	*Duration (Days)*	*Arrival Date*	*Place*
Ridgway	7.5 kn	4 kn (95 mpd)	1 June	295	1 Apr.	7
Blyth	7.5 kn	4 kn (95 mpd)	1 June	295	8 Apr.	8
Knox-Johnston	7.25kn	4.25 kn (108 mpd)	14 June	260	3 Mar.	6
Moitessier	8.5 kn	5 kn (120 mpd)	21 July (P)	234	14 Mar.	5
Fougeron	7 kn	4 kn (95 mpd)	21 July (P)	295	18 May	9
King	9.5 kn	6 kn (144 mpd)	1 Aug. (P)	194	14 Feb.	4
Crowhurst	15 kn	9 kn (220 mpd)	1 Oct. (P)	130	7 Feb.	1
Tetley	15 kn	8 kn (192 mpd)	1 Sept. (P)	146	12 Feb.	3
Howell	15 kn	8 kn (192 mpd)	14 Sept. (P)	146	10 Feb.	2

Crowhurst gave himself a significant edge over Tetley and Tahiti Bill Howell, both racing similar boats and both, according to the chart, with a good head start over him. He might have supposed his revolutionary innovations would provide additional speed, or that he was the tougher man, and would drive his boat harder. Tetley was unknown, but Tahiti Bill's years at sea and tens of thousands of miles as a single-hander had been well-recorded in the yachting press, and Crowhurst's disregard of this was willful.

But the signal fault in his chart was his failure to allow for the wild card of the sea itself, the great leveler that always makes a mockery of men's best-laid plans, and the luck, good and bad, that is inevitably found there for one and all. Crowhurst's chart was the sort that could have been made only by a man who knew nothing of the sea.

Cox Marine delivered the three *Victress* hulls to Eastwoods on schedule on July 28, and the real building of Donald Crowhurst's boat began. A hull (even three of them) is simply a single component of a yacht: it's the platform onto which everything else is added. It is the fastest part to complete. Plywood hulls, such as those for a *Victress* trimaran, can be made in a matter of days by a skilled crew, especially when a boatyard, like Cox, has already produced an identical set of hulls, leaving behind ready-to-reuse molds, templates, and experience. Most of the time and the money spent building a yacht are in the details. This is where Crowhurst's boat bogged down.

John Eastwood spent Sunday, July 28 (the same day the three hulls were delivered to Norfolk), with Donald Crowhurst at his home in Bridgwater. They talked about the details from 9 in the morning until 9 that evening, and Eastwood was impressed by Crowhurst's grasp of the many complex details, his technical facility, and his imaginative mind.

Crowhurst specified a sleeker profile to his boat, without the large cabin Piver had drawn that would be vulnerable to the great sweeping waves of the Southern Ocean. He wanted his boat to be flush-decked with only a small, rounded doghouse immediately forward of the cockpit—a very sensible and seamanlike arrangement, and faster and cheaper to build. He proposed a number of ideas and modifications for his boat that seemed, to Eastwood, well-considered and—some of them—brilliant.

Among them, the cornerstone of his innovative scheme for trimarans, was his system for preventing a capsize: if the boat heeled over dangerously far, electrodes in the side of the hull would send a signal to a switching mechanism (Crowhurst called it his "computer"), which would fire off a carbon dioxide cylinder connected to a pipe inside the hollow mast, which would inflate a buoyancy bag at the top of the mast. This would prevent the boat from completely capsizing. At that point, partially inverted, Crowhurst would pump water into the upper hull, which, as it grew heavier, would push downward and eventually flip the trimaran back upright.

His onboard "computer" would do other things too. It would electronically monitor stresses in the rigging, sounding warning lights and alarms if loads became critical. Hooked up to a wind-speed indicator, it would automatically ease sheets and sails.

The wiring and pipes for carbon dioxide were among the features that John Eastwood was to incorporate into the boat. Crowhurst told Eastwood, and claimed in publicity fliers he was sending out everywhere, that these features had been tested and were "now operating sucessfully" and were the results of a "development project" by Electron Utilisation Ltd.

But no such development project had yet taken place in his workshop in the former stable behind his house in Bridgwater.

A friend described Crowhurst at this time as "grappling with problems and therefore euphoric." He was always the cleverest member of his circle of friends, and his charts and arguments and research must have appeared dazzling to them. But it was not a case of his impressing only those around him, who were not experts, who knew less than he, who lacked Crowhurst's intense authority and ability to amass and overwhelm with knowledge. He had also impressed Angus Primrose, the designer of Bill King's *Galway Blazer* at the London Boat Show in January. He had convinced the pragmatic Stanley Best. He had impressed John Eastwood. Even the *Sunday Times* had reported that he was an "experienced sailor," comparing him favorably with the "rash younger men." Donald Crowhurst had an extraordinary talent for making people believe him. His power lay in the fact that he had completely convinced himself.

9

Six hundred miles south of the equator, 600 miles east of Brazil, beating into the southeast trade winds on the port tack (the boat's port—left—side facing the wind and thus taking the strain), John Ridgway was "horrified," as he wrote in his log, to see the deck bulging around the aft portside shroud plate.

This plate was fixed to the deck by two bolts extending down through the underside of the deck where they were secured with nuts and washers. The flat area of deck was taking all the load from the mast and sails, and showing it. Part of the wire rigging holding the mast up was secured to this plate. This is typical of the lightweight construction used in day-sailers and weekend cruisers. A boat properly designed for ocean voyaging would have its rigging anchored to chain plates—long, thick straps of steel or bronze that are through-bolted to the hull or an interior bulkhead, spreading the loads from the mast downward and through the main structural members of the boat. What Ridgway was looking at now in horror was clear evidence of inadequate construction straining under excessive load. What he feared, quite rightly, was that the shroud plate or the deck could fail and *English Rose* could lose her mast.

He dropped the sails, removed the wire attached to the plate, and unbolted the plate. He replaced it with a new one, through-bolting it through a backing block of plywood on the underside of the deck. This, he hoped, would reinforce the deck around the plate which had shown cracks in the fiberglass gel coat. But when he inspected his repair the next morning he saw that the plywood was now bending with the deck around the plate and was, he recorded, "creaking ominously."

Ridgway still did not grasp the reason for it all. As a sailor he assumed (as do many buyers of yachts) that the experts knew what they were doing; that, knowing of his plan, *English Rose*'s designer and builder had produced a yacht for him that was capable of meeting the conditions he expected to encounter. But they hadn't. Whatever calculations and conclusions the Westerly design and construction team had reached about this specially "beefed-up" boat's suitability for its voyage, they were woefully wrong. Those who should have known better had sent Ridgway off around the world in the hope that his adventurer's pluck and determination and star quality would compensate for their boat's utter unsuitability. They hoped he would somehow simply pull it off. It was like a cooper who knows nothing about waterfalls building a barrel for a man who knows nothing about barrels but insists he can ride one over Niagara Falls.

But Ridgway failed to see this. He continued to believe the problem must be the result of the collision with the trawler.

> I tried to puzzle it out all day. I had photographed the cracks on 1st July. I recorded then, "I do not believe these too ominous." But at the back of my mind I had begun to wonder why there should be cracks on the port side and not the starboard side. The conclusion I reached was that when the trawler hit the starboard bow, on the first day of the voyage, the impact must have cause a sudden "whip" in the mast, which could have strained the plates on the port side. Whatever the cause of the damage, the result had been a steadily increasing bulge in the deck around the chain plate. Had I not replaced this plate on the previous day, I am convinced that it would have pulled

out during the night with disastrous results. The piece of wood only slowed down the inevitable process.

But there had been no damage to the port side; any problems resulting from the whipping of the mast would have become evident on that first day. His guesswork gave him an alternative to what he could not imagine: that *English Rose* was simply showing the signs of being where she manifestly did not belong.

This would have been a grim situation for a cheerful competitor, but to a man who was severely lonely and depressed, it was the end. The "ominous creakings" had done for him. On the evening of July 16, after just over six weeks at sea, John Ridgway gave up. He eased *English Rose* away from her windward slog and headed for Recife, Brazil, the nearest downwind port with a British consulate.

He spent 5 days sailing west in a funk of bitter disappointment with himself, despite the fact that his appetite recovered and he tucked into special treats, like Scotch grouse, packed for holidays and special moments in the race. He thought of all the people who had helped him, whom he now felt he had let down. "I don't think I have ever given up in my life before," he wrote in his logbook. "Now I feel debased and worthless. The future looks empty. . . ."

On July 21 he sailed into Recife, and out of the race.

Ridgway's sponsor, *The People,* would not see its hero defeated by a bit of technical mumbo jumbo. Its headlines back in England provided a nobler exit: "Ridgway Beaten by Mountainous Seas and Gale-Force Winds."

10

In the summery weeks after John Ridgway gave up, it was a quiet race. Chay Blyth's radio was not working (even working, it had poor range and the newspapers reported that his silence was no cause for alarm), and Robin Knox-Johnston, whose powerful Marconi radio was functioning, reported steady but not spectacular progress down the Atlantic. Cricket scores held more people in thrall in England. In the United States, where flower power, the Democratic National Convention in Chicago, and the presidential campaigning of Hubert Humphrey and former Vice President Richard Nixon were news, few people would have been aware of the two Englishmen sailing down the Atlantic in their tiny, unglamorous boats.

In the middle of August, Hurricane Dolly swerved north of the Azores and headed for Europe. Bernard Moitessier and Loïck Fougeron—the "dry, calm French single-handers" the *Sunday Times* had called them when chiding the rash early starters—and Bill King were all ready to leave Plymouth and join the race. But Dolly was producing winds in the English Channel stronger than the three sailors cared for at the start of their voyage. They remained in port, getting regular weather updates from the meteorological office at the Mount Batten

RAF station near Plymouth, doing last-minute jobs on their yachts and growing antsy.

The *Sunday Times* had wanted to give Moitessier a radio so he could transmit reports to them. He refused, on the grounds of preserving his peace of mind, but he did accept a Nikonos camera and dozens of rolls of film in screw-top aluminum canisters: these he could pack with exposed film and messages and shoot onto the decks of passing ships with his slingshot. He had mastered the slingshot as a boy in Vietnam and he told the newsmen that a good slingshot was worth all the transmitters in the world. The pack of yachting journalists hovering over every moment of the Golden Globe racers' preparations enjoyed Moitessier. They photographed him demonstrating his slingshot technique and printed his ascetic harangues: "This is not for making money—screw the money. . . . Money is all right as long as you have enough for a cup of tea. I don't care for it any more than that." They described him as "thin as a streak and brown as a brazil nut" and reported that while he was waiting for a break in the weather, he hoped to see Disney's *The Jungle Book.*

The BBC's shipping forecast on the morning of Thursday, August 22, called for favorable winds for the next two days, but also fog. In Plymouth, however, it was sunny and the weather looked good. Moitessier and Fougeron, their boats anchored in the harbor, called to each other across the water from their moorings, discussed the forecast, and decided to risk the fog rather than wait until Saturday and have the wind turn against them. And they wouldn't depart on a Friday. "Sailors do not like to leave on Friday," wrote Moitessier, "even if they are not superstitious."

Bill King didn't like the possibility of fog. He told the Frenchmen he would wait until Saturday.

Moitessier's wife, Françoise, cried as he raised sail, and he was brusque with her. "Listen, we'll be seeing each other again soon! After all, what is eight or nine months in a lifetime? Don't give me the blues at a time like this!"

She was taken off *Joshua* in a press launch, which then followed the two French boats out of the harbor. Her wifely feelings didn't fit Moitessier's mood of tension and exhilaration. He was anxious to be off.

> I felt such a need to rediscover the wind of the high sea, nothing else counted at that moment. . . . All *Joshua* and I wanted was to be left alone with ourselves. . . . You do not ask a tame seagull why it needs to disappear from time to time toward the open sea. It goes, that's all.

It may have been exciting being Bernard Moitessier's wife, but it was never to be a full-time position, as Françoise must surely have realized by then.

The two Frenchmen sailed close-hauled down Plymouth Sound in a light southeasterly. *Joshua* steered herself as Moitessier, wearing only swimming trunks, stood on deck adjusting sails and rolling a cigarette. Fougeron brought his Moroccan kitten, Roulis, on deck for the press photographers in the launch. Once past the breakwater they eased sheets and headed down-channel. Then *Joshua* pulled ahead.

Moitessier had spent months going over *Joshua*'s gear and rigging, renewing everything necessary for what he considered would be the greatest sailing adventure of his life. Coming back from Tahiti with Françoise, the boat had been loaded with all the equipment and spares two people would take for an extended voyage through the tropics, as well as the junkyard of conceivably useful spare parts, pieces of timber, miscellaneous hardware, and lengths of metal that all boats seem to collect over the years. In Toulon he had offloaded most of it, and later in Plymouth he had further lightened the boat by removing the engine, his anchor windlass, the dinghy, a suitcase full of books, charts, and sailing directions for places he did not expect to go, four anchors, 900 pounds of anchor chain, much of his spare cordage, and 275 pounds of paint in cans. He kept aboard a minimal two anchors, 200 feet of chain, and a coil of nylon anchor line that would stretch his anchoring capability to the deepest of harbors. Also, he rearranged the gear inside the boat better than before, keeping the bow and stern light and buoyant and the heaviest items low down, near the center of the boat.

Where Robin Knox-Johnston had packed aboard *Suhaili* everything he thought he might conceivably use, Moitessier stripped his boat of all but the basics. These two men, probably the most experienced seamen of all the eventual Golden Globe competitors, adopted a fundamentally different approach to loading their vessels. As the race developed the choice was to prove perfectly correct for each.

Lighter and in better trim than she had ever been, with her skipper near the peak of his skill, *Joshua* flew as she never had before. In his first week at sea, Moitessier averaged nearly 150 miles per day, a tremendous speed for a 40-foot monohull, and more than twice as fast as Knox-Johnston's first week average of 71 miles per day. It had taken Knox-Johnston 26 days to reach the Cape Verde Islands off the coast of West Africa; Moitessier was there on September 8, just 17 days out of England. When eventually these times became known in England, it was clear that Moitessier, despite his later start, had every chance of catching up to and overtaking the two front runners, Blyth and Knox-Johnston.

Late into his preparations, Commander Bill King had found that the project was costing him far more than he had anticipated—as can happen with most things, but with boats in particular. He quickly had to find another £7,000. He got a few hundred selling film and television rights; a publisher advanced him money for a book about his voyage. But he was still far short. So he plundered his life savings, sold all his cattle and sheep, leased the grazing rights on his farm, and sold his car.

When he had first thought of his voyage, it was something he was doing for himself; he hadn't seen it as a race, which had come as an unwelcome surprise. But the staking of all his financial resources on his effort now left him with a feeling that would be common to all his rivals, each of whom eventually found every aspect of his life taken over by this race: he had to win.

King sailed from Plymouth on Saturday morning, August 24, two days after Fougeron and Moitessier. *Galway Blazer II* ghosted out of the harbor with an honorary escort of three naval vessels; a navy cannon on the breakwater fired a salute. The sea was quiet in the channel, but the light wind had turned, as Moitessier and Fougeron had feared it would, and was now blowing from the southwest, dead ahead of King. The junk-rig was at its best with the wind on the beam or from aft—the predominant conditions expected on this circumnavigation, and for which King's boat had been expressly designed—but it could not point nearly as close to a headwind, especially a light one, as conventionally rigged modern yachts. He began tacking down-channel, off to a slow start.

With just over two months' headstart, Robin Knox-Johnston was then at 33 degrees south, 13 degrees west, or about 1,500 miles west and a little north of Cape Town, South Africa. He was well south of the tropics now; the weather was growing colder and the wind stronger. He was approaching the Southern Ocean, which officially began at 40 degrees south, just 420 miles ahead. But *Suhaili* was showing more signs of wear, and this was beginning to sap Knox-Johnston's confidence.

The halyard winch brakes were failing: while raising or reefing the mainsail and headsails, the brakes were letting go and dropping the sails onto Knox-Johnston's head. This would be no more than annoying in fine weather, but in strong winds, which had already arrived, it could make sail-handling impossible.

He fixed the brakes temporarily, but on August 6 he noticed that the main gooseneck—the hingelike piece of hardware connecting the main boom to the mast—was beginning to come apart. Failure of the gooseneck would leave him without effective use of his mainsail, which would slow him drastically; coming at the wrong moment it could result in damage that would force him out of the race.

That night, Knox-Johnston wrote in his log that during the day he had seriously considered giving up and making for Cape Town. He tried to cheer himself up by singing along with a Gilbert and Sullivan operetta on his tape recorder and imagining his sea heroes, Drake, Frobisher, and Nelson, looking down on him.

But heroes were not much good as company. Like everyone else in the race—including, occasionally, Bernard Moitessier—Knox-Johnston suffered from loneliness. The first minutes of his voyage, when the boat carrying his family turned back to Falmouth, had been devastating. Two months later, on a Saturday night, listening to Lourenço Marques radio from South Africa, he wrote in his log: "I feel lonely tonight. Listening to L.M. has brought back memories of South Africa. . . . I can remember the parties all too well."

It was his own considerable dogged ingenuity, his rising to challenges, that sustained him. After a radio exchange with a South African station, he found that his battery charger was not charging, and he took it apart. When he'd cleaned grease off the spark plug points, he realized he had no feeler gauge aboard to reset the spark plug gaps. He made his own by counting the pages of his logbook: the gap needed to be between 12/1000 and 15/1000 of an inch. 200 pages to the inch meant that each page was five-thousandths of an inch thick. He set the spark plug gap of 12-15/1000 using three pages and the charger worked. There was a way, it seemed, around any difficulty.

As he neared the Southern Ocean, Knox-Johnston began preparing for the storms he knew were coming. He stored most deck gear below; storm sails, sea anchor, spare lines and lashings were placed ready for quick deployment. From the bulk stores in the forward cabin, he topped up the frequent-use containers of kerosene (light and cooking) and gasoline (battery-charging) that he kept in the saloon. He put away his tropical clothing and got out sweaters, jeans, and socks.

Finally, on August 27, he encountered his first real gale of the voyage. But the wind was not westerly, which he could have expected so close to the Roaring Forties. It blew from the southeast, the direction in which he was trying to go.

Suhaili made little headway against it. She was not designed to pound to windward, and Knox-Johnston reefed her down to avoid straining her. For a night and a day she bobbed alternately northeast and southwest, either side of the eye of the wind, on whichever tack appeared the best course. The seas built up until they were steep and breaking, but the heavy double-ender was in her element. Perfectly balanced under shortened rig, she rode the waves beautifully.

But her Indian-carpentered hatches leaked copiously. Such leaks do not bring the disquiet that comes with water seeping in through the hull below the waterline, but they produce a sodden misery for the sailor in his home at sea. With every wave that broke over the deck and cabin, salt water poured in through the companionway hatch and splashed over the chart table, the book rack, and the Marconi radio. Knox-Johnston covered these with towels and rags, but they only ensured continual dampness. The skylight dripped incessantly above his sleeping bag, which he tried to keep from being soaked by placing a piece of canvas over it. There was so much water coming in that during this first gale Knox-Johnston wore his oilskin jacket and pants below, and these, streaming salt water, added to the thick damp in the cabin.

A small boat at sea is its crew's only port in a storm, and if the boat is cold and wet below, its gear beginning to fail, the dismalness of such a situation can't be exagerrated. It undermines the sailor's most important illusion: that he is safe. As he lay soaked and battered a thousand miles west of Cape Town, the proximity of safety, warmth, and company ate at Knox-Johnston and he thought again of giving up.

Only a few hundred miles away—two days ahead of Knox-Johnston and leading the race—Chay Blyth was battling the same gale. The weather system (traveling across the ocean from the west) that had passed over Robin Knox-Johnston on August 27, reached Blyth and *Dytiscus III,* a lesser sailor and a lesser boat, in the dark early hours of August 28. But nobody was the equal of Chay Blyth for his soldierly attack at whatever was thrown at him.

For two days the gale drove him (as it did Knox-Johnston) northeast, then southwest. When it was over, after a day of respite, gale-force winds rose again from the south. Blyth headed east, pushing *Dytiscus III* as hard as he knew how—harder than he knew he should, for he was now well aware of his boat's frailty. The little weekend cruiser was leaking badly, its gear was showing signs of wear, things were breaking. But what he had been able to do with it so far was remarkable. Knox-Johnston was being careful with *Suhaili,* determined not to strain her during this first gale, but Chay Blyth was actually doing his best to push *Dytiscus III* to the breaking point. South Africa was his last possible stop if he was to abandon the race before heading east and further south into the truly dangerous seas of the Roaring Forties. To break something important, to lose a mast for example, far out in the empty wastes of the Southern Ocean, could be a fatal exercise. If such a catastrophe was to happen, he wanted it sooner rather than later.

On September 6, Chay Blyth sailed into the Roaring Forties.

The next day, his ninety-second at sea, his voyage became as long as his Atlantic row, and as if on cue, signaling Blyth's move into new territory, the servo blade (which served as the trim tab) of his wind vane steering gear broke.

This was the sort of crucial gear failure he had feared, and even expected. But when it came, it threw Blyth into a crisis of indecision. Unlike deeper-keeled boats, *Dytiscus III* did not track well with her shallow bilge keels without constant attention to her helm, whether from Blyth or the wind vane gear. Without self-steering, he could not go on. He replaced the broken blade with his one spare and now thought hard about putting in to South Africa so a new one could be flown out to him.

This would not be allowed by the *Sunday Times,* but this no longer worried him. According to the rules, he had already disqualified himself.

Weeks earlier, he had found that his supply of gasoline had turned a milky white: it had become contaminated with salt water from one of the boat's many leaks. This meant he could no longer run his battery charger, which he relied on for his boat's navigation lights, but more importantly, for his communicating radio. He was immediately concerned that without radio transmissions his wife Maureen would worry about him. He first thought of stopping at South Africa for gas, where of course he would be disqualified. Finally he decided to head for Tristan da Cunha, a small, isolated South Atlantic island group belonging to Britain, where he hoped to pass close enough to shout a message at someone to tell people in England that he was all right.

He reached Tristan on August 15, approaching the high, forbidding island in circumstances that would frighten any sailor. The wind was behind him, a shore of cliffs rose ahead of him, and he had no chart of the sea bottom beneath him—a recipe for disaster. Getting closer, he saw a ship anchored ahead, and he fired off a flare, a distress signal. This is not a seamanlike practice when there is no emergency, but Blyth got the reaction he wanted: a small boat put out from the anchored ship and approached him. The vessel, the men in the boat told him, was the Tolkienesque-sounding *Gillian Gaggins,* and she had arrived from Cape Town only that morning on one of her three annual visits to pump gasoline ashore to the island. The men from the *Gaggins* told him there was nowhere safe to anchor nearby and that he could lie to the ship's stern on a line. Blyth sailed *Dytiscus III* up to the ship and was hailed by its captain, Neil MacAlister, another Scotsman, who invited Blyth aboard for a drink and offered to give him gasoline and whatever assistance he needed. This was too much for Chay Blyth. It bore all the signs of holy intervention.

> It was all entirely unbelievable. . . . My arrival had coincided with [MacAlister's], in the face of odds of about a hundred and twenty to one against, and here he was with a ship which could give me the very stuff I was looking for. . . . This was more than luck. Coincidence alone could not take the credit. God had been a reality to me throughout the trip, and now, yet again, He had manifested His presence.

He wasn't going to question God. God had spared him the decision of whether or not to take on gas by placing a tanker in his path.

For the first time in 9 weeks he stepped off *Dytiscus III* and boarded the *Gillian Gaggins*. He drank good Scotch whiskey with the captain while the ship's engineer fixed his saltwater-impaired generator. He learned for the first time, from Captain MacAlister, that John Ridgway had dropped out of the race. Blyth was stunned. Ridgway, his superior officer who had called the shots through many adventures, had always appeared to him the expert, the real seaman of the two, the better planner, the better-equipped—the better man. Until now he had assumed that his old partner was somewhere ahead of him, that in much the same way they had crossed the Atlantic together, they were still sharing this experience. It had been a comfort. With Ridgway out, Blyth felt suddenly vulnerable. He was in the lead now, on his own, heading toward the Southern Ocean with no one in front of him.

He accepted Captain MacAlister's help and hospitality. He had a hot shower and spent the day aboard the ship sending telegrams to Maureen and others who had helped him with the boat, asking their advice on various problems. He ate dinner and did not resist the invitation to spend the night aboard the *Gillian Gaggins*.

Blyth was torn about what the conditions of the race meant to him. In the telegrams he sent to England from aboard the ship, he said that he was taking aboard fuel—which he knew by then was not allowed under the rules—but that he had not stepped ashore. His focus and interest had shifted from the broader view of the *Sunday Times'* race to his own personal reasons for attempting a nonstop circumnavigation. He had begun asking himself the question they all asked sooner or later, in the face of unrelenting hardship and loneliness: *What am I doing here? What's the point?* In time, each sailor came up with a different answer and acted accordingly. "We're going to win!" he had said to Ridgway at the low point in their canoe race, that goal overriding any other consideration. But Blyth now decided he was far more interested in his own ability to go the distance than in winning the race.

The next morning, while he was eating breakfast aboard the *Gillian Gaggins,* the ship's bosun announced that *Dytiscus III,* which had been lying astern of the ship all night, had broken free and was drifting toward the rocky shore. The wind and sea had risen during the night, forcing the ship to interrupt its pumping operations, and it had steamed farther offshore. The mooring line holding *Dytiscus III* had parted. Blyth was frantic. But MacAlister steamed around the drifting yacht; a grappling hook was thrown, and the boat was taken in tow again. Blyth was carried out to it in the ship's lifeboat, together with cans of gasoline. He raised sail and sped away toward the rough, empty ocean, alone again, but now on his own terms.

Now, three weeks later, riding out another gale, 400 miles from South Africa and 4,400 from Australia, Chay Blyth decided that if he was to go on, he must stop and get spare servo blades. He radioed telegrams to England, with the help of a relaying ship somewhere not far away, asking for spare blades, bolts, and drills to be sent to him at Port Elizabeth. Then he headed for land.

> What it amounted to was that the thing I was still curious about was me. The boat, simply because I had taken it where it had never been intended to go, had failed. . . . But I still did not

> know if I myself could stand up to the circumnavigation—and if I could find out, then I wanted to do so.
>
> This business of making myself thoroughly unpleasant to the body which God gave me is something that has fascinated me for almost as long as I can remember. . . . I cannot say that I enjoyed my Arctic and desert survival courses or the rough parts of the trans-Atlantic crossing any more than I can say I was enjoying having the stuffing knocked out of me in *Dytiscus III*—and yet there is an enjoyment. . . . And I did not want, if I could possibly help it, to miss finding out all I could about this round-the-world exercise simply because my boat was not able to do the thing in one go. Survival, after all, was the object with which I began my preparations, long before a newspaper came along and turned it into a race. Provided I could go on without being foolhardy, I wanted to see the thing through. It was my voyage of discovery, and what I wanted to discover was me.

With the last sentence Blyth stumbles across the credo of all adventurers, be they sailors, mountaineers, or explorers. The where and how is simply the means to burrow as deeply as posssible into oneself. It's the answer to the relentless question that floods the mind when the exercise becomes painful and severe: *What am I doing here? What's the point?*

Another gale blew him east past Port Elizabeth. In the early hours of Friday, September 13, Blyth found himself off East London. He called the port radio station and asked for a tow to come out to pull him in. At 8:30 A.M. the East London pilot launch appeared, threw him a line, and towed him into the harbor. A tow is almost always an ignominious surrender of the sort of self-sufficiency sailors go to sea to find, but tied up at the dock, Blyth would not get off his boat. He still intended to go on; he planned to wait aboard his boat at the dock for his spares to be forwarded to him from Port Elizabeth, and then set off. But his concept of "alone" and "nonstop" had strayed so far from the idea of staying out at sea, that his rationalizations for continuing his voyage, and yet staying aboard his boat while docked, grew strained.

Chick Gough, a former paratrooper friend who had moved to South Africa, appeared on the dock with a bottle of whiskey. He came aboard and they drank the whiskey and Gough went off and came back with some beer. The men drank into the night. After four days at the dock, during which Blyth drank and caroused with Gough and spoke with people ashore, yet never got off his boat, the spare parts arrived. Blyth took them aboard and sailed.

Two days later, in a strong gale—Blyth estimated the wind gusting at around 60 knots and described the seas as "colossal," the biggest he had ever seen—he gave up and turned the boat back to Port Elizabeth. It was the wrong boat, he accepted finally, and he had no business being where he was in it. He telegrammed Maureen what he was doing and asked her to fly out and sail the boat home with him, and she did.

For a neophyte—or for any kind of sailor—he had put up a remarkable performance. In three months of racing, Robin Knox-Johnston, the ingenious and consummate seaman, who had set out six days after him, in a larger boat, had gained only four days on Blyth in his weekend cruiser.

The race was over for Chay Blyth. But he had discovered for himself the awful and compelling laboratory of the sea. He would go back to it to continue his experiment.

II

THE WATERS OFF CAPE HORN have long held the reputation of being the most fearsome a ship or sailor can pass through. However, South Africa's Cape of Good Hope, at latitude 34 degrees south, receives solid doses of Southern Ocean weather and poses its own peculiar hazards for mariners attempting to round Africa between the Atlantic and Indian Oceans.

Their problem is the Agulhas current, an immensely strong confluence of water that funnels south out of the Indian Ocean through the Mozambique Channel until it is a concentrated arterial torrent quite separate from the cold ocean around it, pumping south and west off the cape coast directly against the prevailing westerly wind systems of the Southern Ocean. When these westerly gales, or the cape's sudden "southerly busters," blow across the Agulhas, wind and water collide, creating heaping turbulence on a scale seen nowhere else on Earth. Extraordinary freak seas and holes develop. Ships hundreds of feet long have literally fallen off giant waves, plunged through the deep hole-like troughs, and kept on going down. The Agulhas phenomenon occurs mainly along the 100-fathom line paralleling the coast, but gyres of warm Agulhas water spin off from the main stream and curl like beckoning fingers far out into the cold Southern Ocean. It's the chaotic changeability of South African waters, the suddenness of their disruption, that is unlike the more consistent conditions off Cape Horn. Many knowledgeable sailors are more afraid of rounding South Africa than the Horn. Here catastrophe can occur with devastating suddenness.

Robin Knox-Johnston sailed into the Roaring Forties on September 3, about 500 miles west-southwest of the Cape of Good Hope. It was near the end of the austral winter. A northerly wind pushed him southeast, deeper into the Southern Ocean. That night, portentously, a vagabond gust tore across *Suhaili* and split the small spinnaker flying in the bow.

Crossing the parallel of 40 degrees south did not automatically mean the weather should suddenly become worse, but Knox-Johnston found himself waiting nervously for it, and wishing that what he knew must come eventually would come soon. For two days the weather was fine, but then his shipboard barometer began a steep fall, auguring stormy conditions. Despite this, September 5 dawned nearly calm, and he set more sail. During the afternoon the wind rose, and at 1700 his first Southern Ocean cold front overtook him. It was sudden and violent. Within minutes, the wind backed from the north into the southwest and began to blow at gale force. Brutal waves of hail bounced off *Suhaili*'s deck and drove into Knox-Johnston's hands and face as he put deep reefs in his mainsail and mizzen and replaced the jib with the tiny, heavily reinforced storm jib. Escaping below, wet, cold, and stung, he took a heavy slug of brandy.

Suhaili ran east before the rising wind. A confused cross sea soon developed, as new waves driven by the southwestly wind toppled against the old swells from the north. However, Knox-Johnston felt that "the Admiral," as he called his wind vane, was working well in the stronger wind, and *Suhaili* seemed to be handling the conditions.

He was apprehensive and spent the evening in his foul-weather clothing, lying on the piece of canvas he kept over his sleeping bag, ready for whatever might happen. *Suhaili*'s cabin—a tight 8-foot by 12-foot box, crammed with books, gear, jugs of food, and fuel—swooped up and down and lurched sickeningly from side to side with the violence of a roller coaster ride. The noise of the gale—waves crashing over the boat, seas rushing along the other side of the planking inches from his head, the wind battering the reefed sails and shaking the rigging, causing the whole boat to shudder violently when it wasn't being slammed by water—kept Knox-Johnston awake. The singular noise of high wind in a boat's rigging during a gale at sea has no counterpart in the landbound world, where overhead electrical, cable, and telephone wires are long and run without great tension. Wind howling through these is low in tone and without multiple atonal chords. *Suhaili* was cobwebbed with 30 or more separate lengths of wire and rope running up its masts, tightened or winched to considerable tension. This malevolent eldritch shrieking was hard to endure. It ate at Knox-Johnston's nerves. It was undeniably the sound of imminent disaster, which grew to seem more likely, and finally inevitable, the longer it continued.

That evening, to ease his nerves the way one might by phoning a friend, Knox-Johnston talked into his tape recorder, attempting to describe the conditions and his feelings. But as the night wore on and nothing happened, he drifted off to sleep.

He was woken by the awful roaring that accompanied the cascade of heavy objects falling on top of him. The kerosene lantern hanging from the cabin roof swung violently and went out. As he tried to get out from under the weight heaped on top of him, he realized in the blackness that his small world had shifted 90 degrees on its axis and he was pinned to the side of the boat, which now lay beneath him.

As he struggled free, *Suhaili* lurched violently back upright and he was thrown across the cabin, together with tools, tins of food, a mass of flying objects. The boat had been slammed over on its side and then righted itself, and as he fought through the dark toward the companionway, all he could see was the vision of what he was certain awaited him on deck: stumps of broken masts, a tangle of wire and torn sails, a long night's fight to stay afloat and alive. He was so certain of this that it was several moments before he would believe that the masts he saw standing were real.

Groping in the dark on deck, he found that one of the Admiral's two vanes was bent over and split on the mizzen rigging wire. He worked his way around the deck, feeling and seeing what he could and found no other obvious damage. As he was doing this another huge wave broke over the boat, throwing it over on its beam again, and Knox-Johnston had to hang on hard to avoid being swept overboard as solid water coursed around him. He felt that *Suhaili* was wrongly aligned in the wild cross seas, and he altered course slightly, adjusting the remaining wind vane, which was still steering the boat. Then he went below.

Water was everywhere, sloshing around his feet. Anxiously he began working the bilge pump, expecting at every moment another knockdown wave. The familiarity of pumping water out of the boat began to calm him. When he got the water below the floorboards, he started cleaning up the cabin.

Everything—books, clothes, fruit, tools, medical supplies—was jumbled together and scattered everywhere. While slowly sorting and restowing what he could, he became aware of water still pouring in around the edge of the cabin with every wave that broke aboard, which was often. Knox-Johnston was alarmed to discover large cracks all around the edge of *Suhaili*'s cabin, and to find that the interior bulkheads had been shifted by the force of the wave that had broken over the boat and thrown it on its side. Most likely the cabin had been damaged not by the wave that crashed into it on its windward side but by the sea on its lee, or underside, as it was thrown over. Sea water can be compared to liquid concrete, and if one imagines a boat falling sideways onto gluey form-fitting concrete, it is the cabin, bolted to the deck, not the much stronger hull, that is structurally most vulnerable. The force of the water, or concrete, will be upward as the boat slams down onto it, tending to push the cabin upward, trying to shear it off the deck.

As he was inspecting the damage, another wave broke over the boat and he felt the whole cabin move; another knockdown could result in the entire cabin being torn off, leaving a 6-foot by 12-foot hole in the deck, which would fill with water and sink *Suhaili* immediately. But there was nothing he could do about it then, in the dark, with the weather preventing any attempt at repair. So Knox-Johnston raised the whiskey bottle to his lips, swallowed a mouthful, wrapped himself in canvas, and—somehow—went to sleep.

When he woke in the morning and looked outside, squalls were racing across the sea, driving hail into the water and turning it milky white. But the storm seemed to be past its peak. He made himself porridge for breakfast, and after a mug of coffee and a cigarette he felt "quite happy." From his generous stores, he gathered a quantity of bolts and screws and spent the entire day reinforcing the cabin top.

Two days after the knockdown, conditions had moderated enough for him to repair the bent wind vane without it flying away, but the waves were still large enough to roll *Suhaili* heavily, and Knox-Johnston, working on the tubular outriggers that held the vanes, was repeatedly immersed in cold seas.

The relative quiet was short-lived. Another gale overtook him, and the daily face of the voyage now changed utterly. In the 80 days before he had reached the Southern Ocean, Knox-Johnston had experienced one true gale. Now they tore past at a rate of one every two days. Usually the wind started to blow from the north, backed suddenly into the west while rising to gale or storm force (wind speed between 34 and 60 knots), then subsided in the southwest and south within a few days. The waves heaped up by these strong and fast-changing winds, colliding into each other from two or more directions, created chaotic and dangerous seas. The normal and safest procedure for a yacht the size of *Suhaili* encountering such weather would be to heave to—stop and ride out the storm under reduced sail—but this would not get Knox-Johnston far. The wind was predominantly westerly, pushing him in the desired easterly direction of his circumnavigation, and if he wanted to win he had to keep sailing.

Driving the boat beyond normally safe limits prevented him from getting adequate rest. He slept fully clothed, wrapped up in canvas, grabbing short naps between jolts that threw him across the cabin. Sleep deprivation and continual bruising knocks quickly took their toll on his morale.

September 9th, 1968

I finally awoke at 1100 having had three hours uninterrupted sleep. . . . We were rolling very heavily and it was difficult to stand inside the cabin, but I managed to heat up some soup. . . . I felt very depressed on getting up . . . I used up a lot of nervous energy last night by leaving the jib up, for what—maybe an extra 20 miles if we're lucky—and what difference does 20 miles make when I have about 20,000 to go?

The future does not look particularly bright . . . sitting here being thrown about for the next 150 days . . . with constant soakings as I have to take in or let out sail, is not an exciting prospect. After four gales my hands are worn and cut about badly and I am aware of my fingers on account of the pain from skin tears and broken fingernails. I have bruises all over from being thrown about. My skin itches from constant chafing with wet clothes, and I forget when I last had a proper wash. . . . I feel altogether mentally and physically exhausted and I've been in the Southern Ocean only a week. It seems years since I gybed to turn east and yet it was only last Tuesday night, not six days, and I have another 150 days of it yet. I shall be a zombie in that time. I feel that I have had enough of sailing for the time being; it's about time I made a port, had a long hot bath, a steak with eggs, peas, and new potatoes, followed by a lemon meringue pie, coffee, Drambuie, and a cigar and then a nice long uninterrupted sleep. . . .

A prisoner in Dartmoor doesn't get hard labor like this; the

> public wouldn't stand for it and he has company, however uncongenial. In addition he gets dry clothing and undisturbed sleep. I wonder how the crime rate would be affected if people were sentenced to sail around the world alone, instead of going to prison. It's ten months' solitary confinement with hard labour. . . .

On September 10 the self-steering trim tab broke. Like Chay Blyth, he carried one spare. Replacing it was not easy. The bottom of the tab fitted into a metal shoe, or bar, protruding aft from the boat's keel, 4 feet beneath the surface. Kneeling on the aft deck, Knox-Johnston could not get the trim tab to drop into the shoe. Finally he stripped off his clothes, fortified himself with a mouthful of brandy, and went overboard. As the boat rose and fell in the waves, submerging him completely in the 50-degree water, Knox-Johnston had to hold on with one hand while trying to slip the trim tab in place—something like threading a giant needle in icy surf.

The failure of the trim tab worried him, just as it had Blyth. It had broken after 8,000 miles, but only one week into the Southern Ocean leg of his voyage. He had at least 20 more weeks before rounding the Horn and turning north back into the Atlantic.

A greater concern now revealed itself. To conserve his water supplies, Knox-Johnston had been using the water from his plastic containers in the cabin, refilling them with rainwater he had been able to catch in a bucket hanging from the gooseneck below the mainsail. He had not yet touched his main water supply of 86 gallons in two tanks beneath the cabin sole. Now, to lighten *Suhaili*'s bow, which he felt would help her sail better running downwind in the stronger conditions of the Southern Ocean, he decided to start using the water from the forward tank. When he hooked this tank up to his galley pump, nothing but a brown stinking liquid came out. The tank had become contaminated with seawater, probably a result of the flooding below after his knockdown. He tried the second and larger tank and found the same.

Knox-Johnston lit a cigarette and thought things over. A seaman's dilemma from ancient times: "Water, water, everywhere, Nor any drop to drink." There was no immediate concern: he had 10 gallons left in the plastic containers, more than enough to reach Cape Town if necessary. Could he go on? The 10 gallons was 40 days' water at his present rate of consumption; Australia was about 50 days away. Clearly, with the rain and hail from Southern Ocean gales, he could get that far. He had hundreds of tins of fruit juice. He decided to keep going and hoped that his water-catching kept pace.

On September 13, he learned over the radio that Chay Blyth had stopped at East London and pulled out of the race. He had wondered how close he and Blyth were to each other during the first big gale, and now knew that Blyth had still been two days ahead of him. Chay Blyth was far less a beginner by his voyage's end, and it's interesting to speculate how these two unusually tough young men would have raced against each other if Blyth had sailed a more suitable boat. Knox-Johnston, who did not know the degree of Blyth's inexperience, was sorry to lose the pressure of a close competitor. (He would find it again, in spades.)

He had intended to sail west following the fortieth parallel as closely as possible to avoid some, if not all, of the worst of the Southern Ocean weather. He had been pushed south of this latitude, and now he sailed northeast for a few days. The weather accordingly became better, and he started repairing the split spinnaker in hopes of finding conditions light enough to use it. He tied the sail's boltrope at opposite ends of the cabin to hold it up and taut and began sewing his way along the rope.

> Coming to the end of a length of twine, I used my teeth to help tie a knot, and then tried to stand up. I had not moved more than three inches when I felt a painful wrench . . . my moustache was firmly tied to the spinnaker. . . . I tried to stretch to the nearest point where the rope was made fast, but I was a good foot short. I rolled my eyes round looking for the knife

> but it was tantalizingly out of reach. . . . I could not undo or cut the knot that held me and it was getting on towards beer time. There was only one way out of it. I closed my eyes, gritted my teeth and jerked my head sharply back, tearing myself free. It hurt like hell and tears filled my eyes, but it soon passed off and at least, as I rushed to the mirror to reassure myself, the symmetry of the moustache was not badly upset.

A few days later he had a more serious accident. While he was crouched over the batteries in the cramped "engine room," checking them with a hydrometer, *Suhaili* broached before a large wave. Knox-Johnston fell and battery acid splashed into his left eye. He got to the deck as fast as possible and threw seawater into his eye for five minutes. Then he went below and washed the eye with some precious freshwater. He added eye drops, but the eye was now stinging painfully.

That night he wondered gloomily whether he would lose the sight in his left eye. He thought about turning around and heading for Durban and medical attention. But then he thought about the two formidable competitors, Bill King and Bernard Moitessier, far behind him but no doubt gaining fast. He was leading the race now, and he decided that winning would be worth an eye.

But he hurt!

> *September 22nd, 1968—Day 100*
> *Last Day of Southern Winter.*
>
> Awoke to find us heading north so got up and gybed. I banged my elbow badly during the night and what with that, numerous other bruises, and an eye that throbbed, I felt as if I had just gone through ten rounds with Cassius Clay.

12

RODNEY HALLWORTH WAS THE OWNER of the Devon News Agency. He was an independent gatherer of local news for West Country newspapers, a stringer for the national papers, and a publicist. He had been a crime reporter for the *Daily Mail* and the *Daily Express* newspapers in London, and when he moved to the rather less sensational coastal town of Teignmouth in Devon, and became its public relations officer, he brought his gift for melodramatic spin with him. Far more than is common among journalists—supposedly impartial fly-on-the-wall observers—Hallworth gave the whole of his passionate and emotional nature to his stories and clients.

He first heard of Donald Crowhurst when the *Sunday Times* commissioned him to photograph the mystery entrant in the Golden Globe race. Hallworth sent a photographer to Crowhurst's hometown of Bridgwater, some distance away in Somerset. On his return, the photographer told Hallworth about the electronics wizard who was going to sail his computer-operated trimaran around the world, and mentioned that he didn't have a publicist. Hallworth immediately got in touch with Crowhurst and they met in a hotel in Taunton, a town not far from Bridgwater.

Hallworth, a large, flamboyantly dressed man who exuded

an inescapable bonhomie, found Crowhurst reserved at first. But after a meal together he believed they were great friends. His worldliness, he felt, appealed to Donald Crowhurst, whom he recognized as a man, like himself, who could not be contained by a small provincial town.

Rodney Hallworth seemed heaven-sent to Crowhurst. Stanley Best had agreed to underwrite the basic cost of his boat, on the understanding that Crowhurst would continue to seek and find other sponsors to pay for the additional expenses incurred by the whole enterprise—food for the voyage, navigational gear, Crowhurst's "revolutionary" systems for his boat, and the support of his wife Clare and their four children while he was away. But he had found almost nothing. His many letters sent to companies around Britain publicizing his voyage, his prospects, his own company and its innovative designs, had resulted in little more than ten cases of Heinz tinned food and some Whitbread's barley wine.* To safeguard his investment in the boat and to see that Crowhurst actually got as far as the starting line, Stanley Best found himself forced to dig deeper and pay for many of these expenses. To cover himself, Best gave Crowhurst a second mortgage loan on his house, a situation ensuring that unless he won, and won big, Crowhurst would be ruined.

Hallworth's enthusiasm, his expansive confidence, his promise of press coverage, and his belief that he could attract sponsors, were convincing—he was, like Donald Crowhurst, a man with a gift for convincing others of what he wanted to believe. He proposed that if Crowhurst would start his voyage at Teignmouth and use the name of the town in the boat's name, he

---

*Whitbread's mild interest in Crowhurst's circumnavigation would later grow into something much larger when the brewery became the sponsor of a major international sailing marathon, the Whitbread Round-the-World Race, the sailing world's grand prix event, featuring a fleet of fully crewed maxi sailing yachts, which took place every four years through the late the 1970s, the 1980s, and the early 1990s.

would start a campaign to find sponsors. Crowhurst, eager to comply, suggested *Electron of Teignmouth* to publicize his company. Rodney Hallworth, his main client's most grandiloquent advocate, insisted on *Teignmouth Electron.*

The boat's launch date, August 31, came and went with the trimaran unfinished. A revised deadline of September 12 passed. Eastwoods was bogged down with the extra work specified by Crowhurst: strengthening of the basic Piver design to stand up to the long passage through the Southern Ocean and modifications made for his self-righting apparatus and other equipment. Because of the heavy weight of the inflatable buoyancy bag that was to be lashed to the top of the mainmast, the mast had to be shorter, and this meant a redesign of the entire rigging plan, which Crowhurst had promised—and failed—to supply to the builders. Eastwood and his partner, John Elliot, frequently needed Crowhurst's presence at the yard to help them determine the arrangement of these and many other changes, but the boatyard in Norfolk was on the far side of southern England, then a long day's drive from Bridgwater, and they didn't see enough of him. He was busy taking radio-telegraphy courses in Bristol, working out financial arrangements with Stanley Best and with Clare for his coming absence, arranging for a friend to take over the sale of his Navicators, looking for sponsors, and, perhaps, still tinkering with his "computer" and its much-vaunted systems.

On September 21—two days before the next "without fail" launch date of September 23—Crowhurst and the builders had an angry argument over the phone. Eastwoods told him they were not intending to cover the boat's plywood decks with fiberglass, as the plywood hulls (built by Cox Marine) had been. This fiberglass sheathing of the boat was part of the design's specifications, and essential for preserving the watertightness and structural integrity of the deck. Eastwoods claimed that the delay in determining the rigging plan (John Eastwood had finally drawn up the new plan himself) had left them no time to glass the decks. They were planning simply to paint the bare plywood now, arguing that since the fiberglass was so thin, good polyurethane paint would do just as well (an incorrect claim and an unprofessional suggestion). Crowhurst was furious, but there was nothing he could do about it now. It was too late in the day.

Over the years, Crowhurst's intelligence, and the force of his personality, had convinced Clare Crowhurst that he could pull off anything he set his mind to—he always had. The idea for his circumnavigation seemed to be going true to form: Donald had decided he was going to do it, and here it was happening, just as he had planned. But that night, after his phone call with Eastwoods, Crowhurst was so upset that Clare pleaded with him to refuse delivery of the boat, to abandon the project. To her surprise, he listened to her seriously. "I suppose you're right," he said, "but the whole thing has become too important to me. I've got to go through with it, even if I have to build the boat myself on the way round."

It was the only time Clare asked him to give up. Taking her husband at his word, she supported him and did all she could to help him. She did not try to dissuade him again.

On September 15, the *Sunday Times* reported that Chay Blyth had put in to East London and was now out of the race. The article went on to say that the most recent competitor to start out, Commander Bill King, England's submarine ace sailing his race-designed yacht, was the first sailor to overtake one of his rivals, Loïck Fougeron. On the previous Wednesday, King had radioed that he was between the Cape Verde Islands and the west coast of Africa. On the same day, Fougeron had been sighted by a ship 350 miles southwest of the Canary Islands, or 500 miles astern of King. This seemed like an impressive overhaul by the Englishman in his purpose-built schooner after only three weeks at sea. But Fougeron's pace was sedate. King's average of 110 miles per day was still slower than Francis Chichester's 128 miles per day at the beginning of his voyage, the standard against which the Golden Globe racers were continually measured. Moitessier had not been seen since the beginning of September, and his 143 mile per day average was not yet known.

In the same article, the *Sunday Times* took its first serious look at Donald Crowhurst's preparations, reporting on the trimaran's onboard computer and Crowhurst's "patented" self-righting device.

The paper also reported that Lieutenant-Commander Nigel Tetley, taking unpaid leave from the Royal Navy, was about to sail from Plymouth in his trimaran.

Tetley sailed the next day, Monday, September 16. *Victress* carried a banner on her cabin side proclaiming the name of Tetley's sponsor, Music for Pleasure, which had sent him off with a boatload of cassette tapes. Brass band music blared from his wheelhouse speaker as he ran down Plymouth Sound. He was surrounded by the usual flotilla of press boats, which managed not to collide with him, perhaps due to his escort of navy launches, including a commander-in-chief's "barge" carrying a vice admiral. His wife Eve waved from a nearby boat. Tetley noticed that she was beating time to the music, and as the brass band started a soulful tune, he was overcome with sobs. Later, as he passed Eddystone Lighthouse, when the boats had all turned back, he bolstered his mood with the bagpipe music of the Argyll and Sutherland Highlanders and ate smoked trout for lunch.

The northerly wind was aft and Tetley had set twin running headsails held out to catch the wind by poles mounted on the foredeck. This downwind rig had been popularized by British sailor and author Eric Hiscock, who with his wife Susan had by then circumnavigated twice (by way of the tropics and the Panama and Suez canals in a small 30-footer) using this twin-headsail arrangement to faciliate stelf-steering. As *Victress* moved offshore, the wind rose and Tetley decided to take the twins down, but as he was doing this, the wind caught a loose sail, flung it aback, and broke its wooden pole in two. The sail then collapsed into the water with the broken pole, which unhooked itself and floated away.

During the first night an upper-spreader on the mainmast broke loose and banged and whirled on the end of its wire in the rigging. The motion was too violent for Tetley to go up the mast to make a repair, so he tried to steer the boat on a gentler course farther downwind, but found this difficult without the missing pole. Demoralized by these two early mishaps before he was even out of the English Channel, he consoled himself by eating a chicken Eve had roasted for him, while listening to Handel's *Water Music.* In the afternoon the wind moderated and Tetley got the trimaran heading downwind, climbed aloft on short rungs screwed to the wooden mast, and repaired the broken spreader.

The next day, his run of unlucky accidents continued: while clearing seaweed off the trailing log line, he accidentally dropped it overboard. He had one spare line, which he wisely left coiled in the log's box.

That night the wind was squally, at times reaching gale force and blowing from the southwest, so Tetley tacked northwest to keep clear of the rocky, tide-ripped French coast. On the third day the wind blew at gale force and the burdened trimaran slammed into the rising seas, until he hove to for the night.

From the start, Tetley conscientiously noted the music provided by his sponsor. He faithfully recorded in his logbook what he listened to and when. "Later, ominous black clouds appeared ahead, and clad in oilskins, I sat in the wheelhouse ready for the worst, listening to Schubert's *Unfinished Symphony.*"

In addition to standards of the classical repertoire, Music for Pleasure had given him a mix of tapes that could be called eclectic: albums by Russ Conway (a pianist featured on British television variety shows, a man who looked like a handsome bank manager and grinned fixedly while playing popular music favorites); the Mousehole Male Voice Choir (Mousehole, pronounced "Mowzle," is a village in Cornwall); George Formby, a deeply parochial British music hall (later film and television) entertainer, who sang bawdy tunes in a flat, nasally northern accent while playing a ukulele; the Red Army Choir; music from the Greek islands; and meditations in Indian sitar music.

He also recorded much of what he ate, accompanied by his rations of music: lunch of cold chicken, tomatoes, fruit, and smoked cheese; dinner of Chinese-style chicken and beef, onions, beans, tomatoes, mushrooms, and peppers, with half a bottle of Beaujolais; roast duck and a bottle of wine for another supper; scrambled eggs and cod roe for breakfast. These oddly resemble the meals writer Ian Fleming used to set before another naval commander—James Bond.

Tetley was also eating with a purpose: *Victress* was heavily laden when he departed and sluggish in the strong winds he encountered at the beginning of his voyage. He calculated that his consumption of food, water, wine, and fuel lightened the boat by about 10 pounds daily.

Teignmouth Electron was launched into the river Yare at Brundall, Norfolk, on September 23. Clare Crowhurst made a short speech and swung a champagne bottle against the hull. It failed to break. In the seaman's world of attenuated superstition, this was supposed to be unlucky, an ominous failure at the outset of a vessel's life. But John Eastwood told Clare that the same thing had happened to Sheila Chichester at the launch of *Gypsy Moth IV*.

The boat was far from finished. It was still without masts, rigging, sails, and literally hundreds of pieces of hardware inside and out that made it livable and sailable. Eastwoods' crew of boatbuilders labored another week of long days, while John Eastwood and Crowhurst, who was now in the yard full-time, argued constantly and bitterly, both giving conflicting instructions to the workmen.

They also argued about money. Eastwoods claimed that the extra work of Crowhurst's improvements, additions, and innovations had nearly doubled the cost of the boat, and they demanded that he provide a £1,000 releasing fee before he took the boat away from the yard.

The still-unfinished trimaran was pronounced ready to sail—as far as Teignmouth anyway—on October 2. Crowhurst set out expecting to make the trip in three days.

John Eastwood, his partner John Elliot, Crowhurst's friend Peter Beard from Bridgwater, and some boatbuilders from the yard were the boat's crew on the first leg. Setting off down the Yare, through the picturesque Norfolk Broads, *Teignmouth Electron*'s maiden voyage commenced with ill fortune. As they approached a village, the local chain ferry set out across the river ahead of them. An ebb tide was sweeping them on. John Eastwood thought there was plenty of room and depth on either side of the ferry, but Crowhurst was worried that the hulls might snag on the underwater chain, and he ordered the yardmen on the bow to let go the anchor. As the anchor dug in, the tide swung the boat around and it crunched into pilings near the bank, making a hole in the starboard hull. They sailed on to Great Yarmouth on the Norfolk coast, where the yard's men patched the damage and then left with John Eastwood. At 2 A.M., in rain and wind, Crowhurst, Elliot, and Peter Beard sailed out into the black North Sea.

The seas were rough and all three men became seasick, but Crowhurst was the worst affected. He began vomiting repeatedly, but remained either at the helm or below at the chart table, navigating and steering through an evil night. He was in a terrible mood, angry at his own weakness, and short-tempered with the others, but John Elliot found him impressive. "Oddly enough it was watching him then that really convinced me he was a man to sail around the world. He revealed his incredible determination and stubbornness. Once he had decided to do something, neither disaster nor persuasion could deflect him."

With a favorable wind, the trimaran sailed fast down the East Anglian coast, across the Thames estuary, around North Foreland, and along the Kent coast into the English Channel. The boat's progress so far had pleased Crowhurst. But as they passed the South Goodwin light vessel, the wind changed into the west, on the nose, and *Teignmouth Electron* was abruptly stopped.

Multihulls, because of their shallow keels and poor grip on the water, cannot sail as close to the wind as conventional deep-keeled hulls, and their poor windward performance is the major compromise they suffer in return for greater speed and live-aboard comfort. It's an acceptable compromise for ocean voyaging when a greater proportion of winds are expected to be from aft or on the beam. But tacking down the English Channel in stormy autumnal weather against the prevailing westerly winds and racing tides would be an exasperating exercise for the most seasoned multihull sailor. It took *Teignmouth Electron* five hours to cover the 10 miles from the Goodwin light vessel to Dover, and by then Crowhurst was dismayed by his boat's sluggardly performance.

There were good reasons for it: the boat was unweighted by any stores and so had an even more marginal grip on the water; its new sails and rigging were untried; its shortened mast reduced its effective sail area; and Crowhurst had never sailed a trimaran. All sailboats have their idiosyncrasies, strengths, and weaknesses, and their owners must learn, with time, how to get the best out of them.

Crowhurst and his crew experimented with various sail combinations, but soon the tide turned against them and a short time later they found themselves off the Goodwin light vessel again. They tacked far offshore where the tides ran less hard and came to within 3 miles of the French coast, where the wind fell light. Crowhurst used this respite in the weather to try shipping his outboard motor. This meant getting it out of a storage locker beneath the cockpit and sitting it on a bracket at the stern of the port hull. The outboard weighed nearly 100 pounds, and he wanted to do this by himself without help from his crew. It took him an hour, using a block and tackle on the main boom, during which he grew furious. They motored on down the French coast. A little later, while charging batteries with the boat's portable generator, Crowhurst burned his left hand on the generator's exhaust pipe. Clare Crowhurst, who was superstitious, was disturbed when she later saw the burn. It had erased the lifeline on her husband's palm.

For 3 days they tacked between the French and English coasts, finding at each landfall that they were only a few miles beyond their former positions.

One night, midchannel, as John Elliot slept, Peter Beard asked Crowhurst how he was going to get this boat around the world when they were having such a hard time getting it down the English Channel. Crowhurst dismissed the problem, pointing out that the majority of the winds he would encounter would be favorable.

But what if they weren't? Beard persisted.

"Well, one could always shuttle around in the South Atlantic for a few months," said Crowhurst. He drew a rough map in Beard's logbook, circling an area in the South Atlantic between Africa and South America. "There are places out of the shipping lanes where no one would ever spot a boat like this."

Crowhurst laughed.

After 4 days of fruitless effort, Beard and Elliot announced to Crowhurst that they had to go home. Both had commitments ashore. Elliot promised him he would send two of his men from the yard to replace them. Running the outboard, they put into Newhaven on the Sussex coast. Robin Knox-Johnston, who was now approaching Australia, almost halfway around the world, had stopped here more than four months earlier on his way to Falmouth—that trip down the channel from London had taken *Suhaili* six days.

For two days after the replacement crew arrived, they remained stormbound in Newhaven. Then they sailed for two days, getting as far as Wooten Creek on the Isle of Wight before the two boatbuilders abandoned ship—they'd had enough of sailing with Crowhurst. He sailed on single-handed a few miles along the Isle of Wight shore to Cowes, where he found another late entrant to the Golden Globe race, Alex Carozzo, making his final preparations. Known (in Italy) as "Italy's Chichester," Carozzo had been chief mate on a Liberty ship being delivered to a breakers yard in Japan several years earlier when he built a small sloop on the ship's deck. On arrival in Japan, he launched his sailboat over the ship's side and headed for California. Ten days out, the sloop was dismasted in a typhoon. Carozzo spent 83 days sailing to Midway Island, where he repaired his rig and shoved off on a 53 day passage to San Francisco.

Construction on his 66-footer, built expressly for the Golden Globe race, had begun at the Medina Yacht Company in Cowes on August 19, and was largely finished 7 weeks later, an incredible accomplishment. Carozzo's audacity and confidence were a tonic to Crowhurst, who spent a day in Cowes talking with the Italian.

On Sunday, October 13, Crowhurst set out from Cowes with a local sailor, another Royal Navy lieutenant-commander, Peter Eden, as crew. They made four long tacks across the channel before reaching Teignmouth two days later on October 15. Eden was later to report that Crowhurst's sailing skills were good, but that his navigation was "a mite slapdash."

It had taken Crowhurst 13 days to cover the roughly 300 miles from Great Yarmouth to Teignmouth, at an average speed of 23 miles per day.

He now had sixteen days before the *Sunday Times'* October 31 deadline to prepare his boat to sail around the world.

13

On Friday, September 20, Loïck Fougeron came close enough to a fishing boat off the Cape Verde Islands to hand over a plastic bag containing film and letters for the *Sunday Times*. He also handed over his cat, Roulis. She had eaten through the aerial wire of his radio (receiver), chewed through bags containing powdered eggs, and spread the powder around the inside of his boat, she had fleas, but most worryingly, he believed she was pregnant. Fearing a boatload of kittens, he wrote the *Times,* he had even considered making up a raft and setting the cat adrift near land.

The plastic bag and cat were delivered to a Mr. Foulde, the British consul for the Cape Verde Islands. The bag was sent on to the *Sunday Times* in London, but Mr. Foulde kept the cat. Very soon, however, he sent a cable to the newspaper asking to be relieved of the cat, which had proved too destructive to the consulate. The *Sunday Times* was not in the cat business, and eventually Mr. Foulde found a home for Roulis on the island.

Fougeron's progress was slow. He was enjoying himself, and the novelty of a single-handed voyage. But he knew that he had already been overtaken by Bill King (King regularly radioed his position back to England, and Fougeron would have heard this on his receiver), and that his friend Bernard had streaked far ahead of him (Moitessier was then past the equator in the South Atlantic), and any hopes Fougeron might have had of winning the race would have been fading fast.

On September 29, Moitessier raised the remote South Atlantic island of Trindade out of the horizon ahead. He hoped to drop off film and letters for the *Sunday Times* there. The island belonged to Brazil, and his South American pilot book, which contained sailing directions for Trindade, was one of those he had left behind while lightening his boat. Still, the land was high and jagged, and by the look of it he believed there would be deep water inshore, allowing him to approach close enough to attract attention, and perhaps a boat to come out to him.

As he sailed nearer and the island's details grew clear he saw the beautiful green of land that is always startling after weeks at sea. He hoped to get close enough to smell it.

The weather was good, clear and sunny, the wind moderate from the northeast, and *Joshua* glided in from the sea like an albatross. Through binoculars he spotted the roofs of a settlement and the wreck of an old iron ship half-sunk off the village.

He let off a blast from his foghorn and sailed back and forth along the shore. Not a living soul appeared. It was midday on a Sunday, and he wondered if everyone was at church, or indoors having a noisy Sunday lunch. After an hour and repeated blasts from the horn, he was on the point of leaving when people began to pour out of a house and stare back at him, one of them with binoculars. Otherwise, they didn't move. Moitessier raised his MIK signal flags (which, when flown together, mean, "Please report my position to Lloyds of London) hoping the man peering at him through binoculars would read *Joshua*'s big white letters and report him to somebody. Still the people ashore didn't move but watched him as though he were an apparition.

He gybed, waved again with both arms in a gesture meaning

good-bye, and headed *Joshua* out to sea. Suddenly, as if breaking from a spell, all the people ashore began waving madly. Some ran down the beach into the sea up to their waists, shouting after him, imploring him. But they had no boat that could come out to him, and without a chart or sailing directions, Moitessier dared sail in no further. He sailed away.

He sailed southeast into cooler seas, steering now for Cape Town, where he hoped to shoot off a packet of mail. Well north of the fortieth parallel (south), in an area supposedly dominated by easterly winds, he found strong westerlies and big seas, almost Southern Ocean conditions. Moitessier clapped on sail and streaked diagonally across the South Atlantic in a series of tremendous daily runs, as high as 182 miles in one 24-hour period. The wind continued from the west, strong enough to make him reef his sails and send *Joshua* surfing, yet these were conditions with which he and his boat were familiar, and he was happy. In a week he covered 1,112 miles—an average of 158 miles per day, well above the Chichester pace.

Ever since his dramatic storm-surfing episode in the Southern Ocean between Tahiti and Cape Horn, Moitessier had believed that lightness and speed were the key to fast and safe passage-making. Without a doubt, he now realized, the ton of excess weight he had removed from the boat in France and England had improved its performance. Now, in a ruthless, exhilarated mood, he looked around inside the boat and began to throw overboard anything else he could consider deadweight: a box of army biscuits (35 pounds), a case of condensed milk (45 pounds), 25 bottles of wine, 45 pounds of rice, 10 pounds of sugar, 30 pounds of jam, a box of batteries, four jerry cans of kerosene, gallons of denatured alcohol, a coil of 3/4-inch nylon line weighing about 60 pounds. (Thirty-three years ago, in a world apparently far less threatened by pollution than we know it to be today, such a dumping at sea would have seemed harmless, even among enlightened nature lovers like Moitessier.)

This enabled him to empty *Joshua*'s forward and aft cabins completely, concentrating all her weight amidships, making her more buoyant and able to surf in heavy seas. It also meant the boat would sail faster in light airs and require less sail raised to drive it in heavy winds.

Stripping away the last threads of superfluous restraint (and one must suspect that his wife Françoise and some of the tidier notions of marriage had been one of these), Moitessier was getting closer and closer to his idealized state of man and ship flying as one across the sea in a way few had ever approached. He also knew he was becoming a stronger competitor in the race every day. "The great game of the high latitudes is just ahead," he wrote in his logbook. And he and *Joshua* were readier than they had ever been.

On October 19, his fifty-ninth day at sea, Moitessier's noon sight placed him 40 miles southwest of Cape Agulhas (the true southerly promontory of Africa, 30 miles father south than the more famous and picturesque Cape of Good Hope). He had two plastic bags full of film—he had photographed every page of his logbook—that he wanted now to throw aboard some vessel and have sent on to the *Sunday Times.* This closing with land and shipping, and the risks of embayment and collision, went against all his seaman's instincts, which would have him steering clear and flying on south and east into the Southern Ocean. But letting the *Sunday Times* know where he was and how fast he was sailing was suddenly important to him. He had become gripped by the race; he knew he was already making a rare and phenomenal passage and he wanted the world to know it. He also wanted to tell his friends and family that he was well. Pushing his instincts aside, he headed north.

As he did so, the wind changed from the west to the southeast, and from the appearance of the sky, his barometer, and his own beautifully attuned weather sense, Moitessier believed a southeasterly gale was imminent. To make use of the new wind, and to be able perhaps to get away before it grew stronger, he headed *Joshua* up the coast between Cape Agulhas and Cape Town, making for a small port he found on his chart called Walker Bay. There could be yachts there, and, it being Sunday again, he hoped he might find one sailing in the bay onto which he could lob his packets. The yacht might even have news of the locations of his friends in the race, Fougeron, King, and Tetley.

As he neared Walker Bay on Sunday, October 20, the wind rose to about 30 knots but the sky remained cloudless, typical conditions for the Cape's notorious "southerly busters," and he grew convinced that no yachts would be out day-sailing in such weather. Meanwhile a steady stream of tankers and freighters was passing him; his MIK flags were flying and now Moitessier worried that they would all report his position and course—north, apparently Cape Town-bound—to Lloyds, confusing the *Sunday Times* and his friends and family about his intentions.

One small freighter was slowly overhauling him and he saw that it would pass close by his starboard side. Close enough perhaps for him to throw his package aboard. It was a gamble: a ship of any size—another yacht even—can blanket a sailboat's wind and render it unmaneuverable, and a sailor's instinctive fear in such a situation is that the crew of the other ship might misjudge the distance and cause a collision. There is also the visceral, unreasoning, hackle-raising fear engendered by the unreal enormousness of a ship that close at sea, with nothing in the visible world to give it scale but one's own tiny boat. The sailor feels like an ant negotiating space with an elephant. But with the rising wind and deteriorating weather, this one looked like today's best bet. Quickly, Moitessier wrote out a message asking the captain to slow down and remain on a straight course so he could throw over a package. He put this message in a film can weighted with lead, grabbed his slingshot, and waited as the freighter drew up.

> The black freighter is 25 yards off to my right. Three men are watching me from the bridge. Snap!—the message lands on the ship's foredeck. One of the officers twirls a finger at his temple, as if to say I must be a little nuts to be shooting at them. . . . I yell, "Message! Message!" They just stare at me, bug-eyed. At this range, with lead balls, I could knock their three hats off with three shots. . . .

> The bridge is almost beyond us: I have to salvage the situation fast. I brandish the package, and make as if to give it to them. An officer acknowledges with a wave, and puts the helm over to kick the stern my way. In a few seconds, the main deck is 10 or 12 yards off. I toss a package. Perfect!
>
> It is time I pulled away, but I am going to make a serious mistake by throwing the second package instead of racing to the tiller to steer clear. I won up and down the line with my first package; I will lose it all with the second. By the time I dash to the tiller, it is already late. The freighter's stern is still slewing my way. To make matters worse, she has blanketed my sails by passing me to starboard while I was on the starboard tack.
>
> *Joshua* begins to pull clear, but not fast enough. By a hair, the stern's overhang snags the mainmast. There is a horrible noise, and a shower of black paint falls on the deck; the masthead shroud is ripped loose, then the upper spreader shroud. My guts twist into knots. The push on the mast makes *Joshua* heel, she luffs up toward the freighter . . . and wham! the bowsprit is twisted 20 or 25 degrees to port.

The ship changed course as if to come back to help, but he waved "all's well," fearful that if it came back it would finish him off.

Moitessier was stunned, then angry with himself, and finally grateful the damage was not worse: he was amazed that the mainmast had not snapped. At the moment of collision, the solid telephone pole had "looked like a fishing rod bent by a big tuna," and had sprung back straight. The two shrouds, whose ends had only slipped in their cable clamps at deck level, were easily repaired. The spreaders were flexibly mounted on the mast and were straight and undamaged when Moitessier had tightened and reclamped the bottom ends of the shrouds.

The bowsprit—the long spar that extends the rig forward of the bow—was the problem. A wooden bowsprit would have snapped like a dry twig as it hit the freighter's hull, and the loss of the sprit would have seriously reduced the sail area of *Joshua*'s rig, and probably caused Moitessier to drop out of the race. But *Joshua*'s bowsprit was a steel pipe, 3 inches in diameter, with a 3/16-inch-thick wall, 6 feet, 10 inches long. With the marvelous elasticity of steel, it had simply bent, severely. But it was now too bent to use effectively; the forestay, on which the big pulling sails of the rig's fore triangle were hanked, was now canted far off center to the left. It would not be impossible to sail in this condition, but the boat's efficiency would be reduced, and worse, in Moitessier's eyes, *Joshua* was marred, and a stain was thrown across the sharpened mystic beauty of his voyage—no little thing to this man.

The expected southeasterly gale came in the night, and Moitessier spent it hove to, unable to sleep while thinking about the bowsprit. He spent all the next day, Monday, October 21, riding out the gale, trying to figure out what he could do to repair the bowsprit. He remembered what César, the foreman overseeing *Joshua*'s construction in the boiler factory, used to say about steel at moments when it appeared reluctant to take on the shapes required of it: "Man is always the strongest."

Finally, on Tuesday afternoon, with the wind and sea down, he went to work. Fixing a chain to the end of the bowsprit, he ran a four-part block and tackle between the chain and the cockpit winch. With heavy shackles, he attached the spare mizzen boom between the base of the bowsprit and the chain to act as a strut, increasing the angle of purchase. Then he went back to the cockpit and began winding the winch handle. Very slowly, to his amazement and joy, the bowsprit straightened until it was almost as it had been before the accident. He tightened the bowsprit's bobstay and whisker stays and straightened the galvanized steel-tube pulpit in the bow where it had been bent. The boat appeared as good as new.

> Worn out by fatigue and emotion, I fall into bed after swallowing a can of soup for dinner. I am tremendously tired, yet I feel crammed with dynamite, ready to level the whole world and forgive it everything. Today, I played and won. My beautiful boat . . . is as beautiful as ever.

The film cans Moitessier had tossed aboard the freighter—the Greek-registered *Orient Transporter*—quickly reached the *Sunday Times*. The following Sunday, October 20, the newspaper ran an article with photographs of *Joshua* rushing through the sea. The article also reported his position and the last known positions of the other competitors.

Using their speeds to date as the basis for projections, the article ranked the competitors according to who was most likely to pick up the two prizes, the £5,000 for the fastest voyage and the Golden Globe for the first yacht home. It was immediately clear that in both cases Moitessier was the man to beat.

POSITIONS IN GOLDEN GLOBE

Competitor	*Estimated Length of Voyage in Days*	*Estimated Finishing Date (1969)*
Bernard Moitessier	246	April 24
Robin Knox-Johnston	330	May 10
Commander Bill King	336	July 26
Loïck Fougeron	339	July 26
Lt-Cmdr Nigel Tetley	319	August 1

The newspaper still judged that Moitessier's overall time and speed would fall below Chichester's (who had taken 226 days), finding that his average so far was 9 percent slower.

What the article revealed for the first time was that the boats everyone had presumed should prove the fastest were not fulfilling their promise: Nigel Tetley's Piver trimaran had taken 8 days to cover his first 510 miles—that's a laggardly 64 miles per day, slower even than Knox-Johnston's first week in his tubby monohull. And Bill King, in his specially designed racing machine that displaced less than half the weight of the steel

Joshua and should therefore have proved faster, was so far only averaging 110 miles per day, and was ranked behind Robin Knox-Johnston. "It's not the ships but the men in them," the old saw goes, and it seemed to be the case still. But these averages were misleading: Knox-Johnston's daily mileage figure reflected the great increase in his speed since he had reached the high winds of the Southern Ocean; King had actually outperformed him in the early stages. The projections were simply an exercise based on the sketchiest of details for the benefit of the newspaper and the readers it was attempting to interest in "its" race. In the end, the final race result bore no resemblance to the *Sunday Times'* careful calculations, or to anyone's best guess of who might win.

Bill King soon heard over his radio of Moitessier's position and his dazzling runs. The news may not have surprised him, but it robbed him of an elemental requirement for bashing on alone around the world for the better part of a year when one has staked all on such a voyage: a reasonable hope of winning. He tried to shrug this off, unconvincingly, in his logbook (his daily entries were written as letters to his wife, each beginning, "My Darling").

> My Darling . . . This evening I learnt that Bernard Moitessier has worked out a big lead. This, of course, must be a great disappointment to me and destroy my peace of mind. I built this boat specifically to pioneer this trip, not for ocean racing. When it transpired that a race was on, there was nothing else to do but join it, but now I have to realize I have little chance of winning it. Already I face the same sort of emotional situation that must have faced Scott when Amundsen reached the South Pole first. . . . This sort of thing is a test and discipline of one's character which must be faced, but I did not set out to test my character. . . .
>
> I cannot drive *Galway Blazer* against faster boats. I will plod on around the world, revelling in my boats's special poetic

> beauty, in her strength and power. I will put disappointments from me—but I wish now I had no radio contact.

But he could not put away his disappointment. He later wrote:

> I had a great struggle with depression over the slowness of my progress. The peace of the long sail, of the months away from mankind with only the sea and the sky with which to battle, and my beautiful boat as a companion, this peace is wrecked by the nagging knowledge that I am in a race and reluctant to force the pace.

That October 20, the *Sunday Times* also reported that Donald Crowhurst would set sail early in the week.

14

As soon as *Teignmouth Electron* reached its home port, it was hauled out of the water at the Morgan Giles boatyard. Men from the Eastwoods yard had come down from Norfolk to fix the leaking "watertight" hatch on the cockpit floor (beneath which lay the boat's generator) and, essentially, to finish building the boat. Piles of equipment, stores, spares, and donations Rodney Hallworth had secured from local merchants—wedges of cheese and bottles of sherry—began to accumulate on the dock beside the boat. Stanley Best brought in campers to house the Eastwoods, the Elliots, and himself and other Crowhurst friends. All these people turned to the enormous job of attempting to make the boat ready for sea—for a voyage around the world—in a little over two weeks.

From beginning to end, it was chaos. The local fishermen and boatbuilders gathered in their workingman's pub, the Lifeboat, to heap scorn and ridicule upon the man and his boat. Crowhurst, they decided, appeared to be in a "daze," unable to supervise the preparation of his boat, which they described as "a right load of plywood."

Crowhurst's friends also noticed a wholly uncharacteristic, subdued numbness settle over his normally effusive personality.

The enormity of what remained to be done, the incredible range of details, overwhelmed him and scattered the focus of his mind.

He carried around loose sheets of paper that were paradigms of the greater disorganization around him: notes of unfinished work, diagrams made for himself and others, addresses, phone numbers, workmen's hours, half-drafted letters, reminders, lists of things to buy—socks, blowtorch, brass strips, gloves, hacksaw blades, pencils, logbooks, lighter with fuel and flints, barometer, flares, life jacket, screws, bolts, tools. Sometimes he was able to delegate some of these purchases to others. Stanley Best was sent to a used-car dealer to buy two dozen small electrical fuel pumps, though he had no idea what they were for. Clare Crowhurst was sent to a local baker to find a recipe for baking bread at sea.

Despite his promises of local sponsorship, Rodney Hallworth had raised only £250. Yet he would haul Crowhurst off at a moment's notice to attend local functions that might only possibly yield more money but certainly boosted the civic pride of Teignmouth. Crowhurst continued to write letters asking for sponsorship or donations of equipment until just days before he sailed.

Reeling from a vertigo of overwhelming details, he would fix on some item and spend hours tracking it down, even getting into his minivan and driving great distances to find some small thing he wanted. No doubt it was a relief to get away.

A BBC television crew arrived to film the preparations. Crowhurst found time to give them a lengthy interview, during which he sounded like a Cape Horner: "I have felt a community with long dead seamen on many occasions . . . seamen who have come this way centuries before you would understand your feelings, and you understand theirs. . . ."

When asked by the interviewer, Donald Kerr, if he had ever faced a situation at sea when he thought he was going to drown, he recounted, in detail, an episode when he fell overboard while sailing single-handed along the south coast. Fortunately, he said, the boat headed up into the wind after a quarter of a mile and he was able to swim back to it. A quarter-mile swim in the hypothermia-inducing English Channel would be a lucky and heroic accomplishment. He had never mentioned this either to Clare or to any friends, although it was exactly the sort of tale he would have loved to tell.

Five days before he was due to sail, Crowhurst, John Elliot, and a BBC camera crew took *Teignmouth Electron* out for a day of sea trials. The genoa sheet track quickly began lifting from the deck, its screws pulling up through the plywood either because they were too small or had not been sunk into backing blocks below the deck. The gasketing beneath the newly rebuilt hatch in the cockpit floor peeled away when the hatch was raised. Crowhurst spent much of the day angrily complaining to John Elliot about the hardware Eastwoods had installed and trying various sail and sheeting combinations to see if the boat's performance to windward had improved, but it sailed no better than before.

The BBC crew hung around through the last days before Crowhurst's departure. At a certain point, Donald Kerr told the crew to change the emphasis of their coverage. He sensed a tragedy in the making.

On October 30, the day before the *Sunday Times'* departure deadline, *Teignmouth Electron* remained an unfinished project, surrounded by piles of stores and equipment on the dock. Donald Kerr told his camera crew to stop filming and they began to help where they could through the last hours of preparation. They went off into town with lists of things to buy. At teatime—that English break observed by working men with the same sacrosanctity once accorded the Queen's Christmas Day address to her nation—Kerr pulled Donald and Clare away to a local tea shop. Crowhurst was in a grim mood and kept saying, "It's no good. It's no good." It seemed to Kerr that Crowhurst didn't want to go but couldn't bring himself to call it off.

That evening, the Crowhursts ate a last dinner at their hotel, the Royal Hotel, with Clare's sister and Ron Winspear, one of

Donald's best friends. The hotel proprietor gave them a bottle of champagne but the mood was funereal.

After dinner the Eastwoods, Elliots, Beards, Stanley Best, and Rodney Hallworth joined them for a drink. Only the irrepressible Hallworth still saw the Crowhurst of his image-making: "He was cheery and raring to go." Hallworth wanted Miss Teignmouth 1968 to sail aboard *Teignmouth Electron* as far as the starting line, give Crowhurst a kiss, and leap overboard as a gun went off. This didn't happen.

After the drink, Crowhurst and Clare rowed out into the cold late October night, away from the cheery lights of shore, to where the boat was moored in the harbor. Unstowed equipment still lay piled on deck and below. They worked on the boat until two in the morning before returning to the hotel. In bed, Crowhurst was silent. "Darling," he said finally, "I'm very disappointed in the boat. She's not right. I'm not prepared. If I leave with things in this hopeless state, will you go out of your mind with worry?"

Clare bravely did what she thought was best. "If you give up now," she said, "will you be unhappy for the rest of your life?"

Crowhurst didn't answer. He started to cry. He cried all night.

The weather was raw and drizzly in Teignmouth on Thursday, October 31, 1968. It was a miserable day to go to sea.

Crowhurst and his team spent most of the day carrying supplies aboard and rushing around town making last-minute purchases. Rodney Hallworth's more dramatic ideas for a sendoff had been vetoed, but he did inveigle Crowhurst into a nearby chapel, where he hoped to photograph the intrepid lone mariner in an attitude of prayer. Crowhurst, wearing a tie, merely sat in a pew, leaning forward slightly, looking tired and pensive.

Meanwhile, Clare Crowhurst filled a carrier bag with buns, ham, and salad from the Royal Hotel, together with some gifts: a book of yoga exercises, a china spoon, a box of cherry nougat, a ventriloquist's doll—her Christmas present for her husband—and a long letter. She took the carrier bag down to the boat and put it on Crowhurst's bunk.

At 3 P.M.—9 hours before the *Sunday Times* cutoff deadline—*Teignmouth Electron* was towed out of the harbor by the local pilot boat, accompanied by three launches carrying forty friends, well-wishers, reporters, photographers, and Donald Kerr's BBC film crew.

Crowhurst started to raise his sails and immediately had trouble. John Elliot had hanked his jib and staysail onto the wrong stays. Their halyards had been wrapped in the lashings at the top of the mainmast that secured the large, heavy, deflated buoyance bag. Unable to raise sail, *Teignmouth Electron* was ignominiously towed back to shore, to the vocal and guffawing delight of the harbor skeptics.

Back at the dock, a Morgan Giles rigger climbed up the mast to free the halyards while Crowhurst hanked on his headsails in their proper postions. (It is striking that he did not go aloft to free the halyards himself to make sure it was done properly; with so much of his boat's preparation necessarily delegated to others, it's possible he had never been up his boat's mast at all.) Rodney Hallworth, ever ready to forge an association or provide a plug for any Teignmouth body, chose this frantic moment to come aboard and hoist the burgee of the Teignmouth Corinthian Yacht Club to the masthead (where it might easily have fouled the buoyancy bag's lashings).

Towed out again, now with the early autumnal dusk falling, Crowhurst raised sail and the trimaran crossed a locally designated starting line at 4.52 P.M. The Yacht Club fired a gun. The wind was southerly and strong, forcing him to beat to windward (the trimaran's worst point of sail) across Lyme Bay to clear Torquay, Brixham, and the long southerly jut of land that stretched away to Start Point, the southernmost tip of Devon. The motorboats, with Clare in the bow of the pilot boat, followed him only for a mile before he disappeared into the murk of rain and the early onset of night.

Evidence of the disorganized sendoff remained ashore. When he returned to the dock, John Elliot was dismayed to see a load of hardware and spare cuts of plywood shaped for emergency repairs lying on the Morgan Giles slipway. He had personally put these things aboard the trimaran. The Morgan Giles men later said these supplies were never put aboard.

Two days later, Stanley Best appeared at the Crowhursts' house in Bridgwater with the carrier bag full of presents that Clare had left on her husband's bunk. It had been found on the slipway with the other supplies.

S

15

SEVEN GOLDEN GLOBE COMPETITORS now lay scattered across half a world of ocean.

"Italy's Chichester," Alex Carozzo, also officially "sailed" on October 31. He was, if possible, even less ready than Donald Crowhurst—or perhaps more deliberate—and simply moved his boat, *Gancia Americano,* to a mooring off the boatyard at Cowes, on the Isle of Wight, where it had been built. There it would swing in the tide, technically departed, until Carozzo felt he was ready to put to sea. The 66-footer, a cold-molded wooden monohull, was strong and light, and while it was then almost impossible for Carozzo to catch up to the leaders, he seemed a likely candidate to take the £5,000 prize for fastest time. Certainly, he should have worried Donald Crowhurst.

As Crowhurst was tacking south and west through the English Channel, 5,000 miles down the Atlantic, Nigel Tetley was nearing the Brazilian island of Trindade, where Moitessier had waved to the astounded inhabitants almost a month earlier. The wind had been light all day and fell away to nothing at dusk, leaving him becalmed. Tetley's radio had been giving him trouble for a week, and he had not been able to broadcast his position. He was worried that Eve would be anxious about him. He tried Cape Town Radio again that evening as *Victress* lay becalmed, but got no response. "It was a disturbing experience trying to communicate with no one answering," he wrote in his logbook, "as if one were dead."

Eighteen hundred miles to the southeast—roughly midway between South America and South Africa—Loïck Fougeron had a lot more wind.

The day before, October 30, he had passed the island of Tristan da Cunha. Conditions had been light to calm all morning, but in the early afternoon the wind reappeared and within a few hours was blowing at strong gale force. It soon whipped up seas larger than Fougeron had ever seen before. They broke repeatedly over *Captain Browne,* exploding on deck with such force that he was worried the boat's revolving Perspex hatches would be smashed, sending water flooding below. By the early hours of October 31, he estimated the wind at hurricane force. Above the crashing seas, the sky was occasionally clear and a brilliant full moon lit up the wild scene. It was the worst weather he had ever seen, and he was afraid.

Only a few hundred miles away to the southeast, a little ahead of Fougeron, Bill King and *Galway Blazer II* lay under bare poles riding out the same storm.

> I had seen storms from the tiny platform of submarines or on sailing boats all over the world; but no mental picture ever occurred to me of the typhoon tumult which now battered *Galway Blazer.* . . . In the very midst of the hurricane, the sky cleared, and I saw a full moon flaming coldly, detached from the awful scene.

Yet *Galway Blazer* rode the seas well through a long night and a day, and toward the evening of October 31—Halloween night—King felt the worst had passed.

Sailing Route Across the Indian Ocean

Two thousand miles to the east—1,100 miles the other side of Cape Town, well into the Indian Ocean—Bernard Moitessier was sailing through the choppy aftermath of a brief gale that had struck the day before. The wind, which had begun in the southeast, had backed through the northeast to the north during the night, before dropping away to a breeze on the morning of October 31.

But the surface of the sea around *Joshua* was still disturbed by the winds that had swept fast through 180 degrees of conflicting directions. Heaps of foaming water tossed around the boat making a strange noise. It made Moitessier think of the sound of a multitude of termites clicking their mandibles or the scattering of dead leaves. The confusion of heaping waves breaking against other waves in nearly calm conditions astonished him, demonstrating that the sea, no matter how well one knew it, was always capable of showing something new.

The gale had not been severe enough to trouble him, but a few days earlier, in similarly rough, disturbed seas, *Joshua* had been knocked flat. An enormous breaking sea had smashed into the side of the boat, so hard that for a moment Moitessier could not believe his portholes had not shattered. *Joshua* came up quickly and sailed on. The only damage on deck had been the wind vane, smashed when the mizzen boom was pushed across the deck. But the vane was simply constructed and Moitessier quickly shipped a spare.

An hour later, sitting on his perch beneath the turret hatch, he saw a freak wave, twice as high as those around it, rise up astern. He jumped down, wrapped his arms around the chart table, braced his legs, and felt *Joshua* take off in a surge of acceleration. Then she went over again, slammed down on her side.

Once more, the buoyant, boilerplate *Joshua* rose upright. Her telephone pole wire and rigging suffered no damage. The wind vane held the boat on course, and she ran on downwind without a touch from her captain.

Moitessier's *Joshua*. Made of boilerplate, like her skipper.

John Ridgway. He had no love of the sea. The voyage was simply an ordeal to endure.

On departure day, Chay Blyth did not yet know how to sail.

Ridgway aboard *English Rose IV* moments before collision.

The forging of a strong bond. *From left to right:* Tetley, King, Moitessier, Fougeron aboard King's *Galway Blazer II* at Plymouth.
Rivalry was put aside and the sailors became close friends.

Robin Knox-Johnston.
A psychiatrist who saw him before he sailed found him "distressingly normal."

Knox-Johnston's *Suhaili* leaving London, low in the water with a year's supply of corned beef.

Knox-Johnston on *Suhaili*'s bowsprit. He didn't like to wear a saftey harness.

Bernard Moitessier.
He went to sea to save his soul,
but found himself in a race.

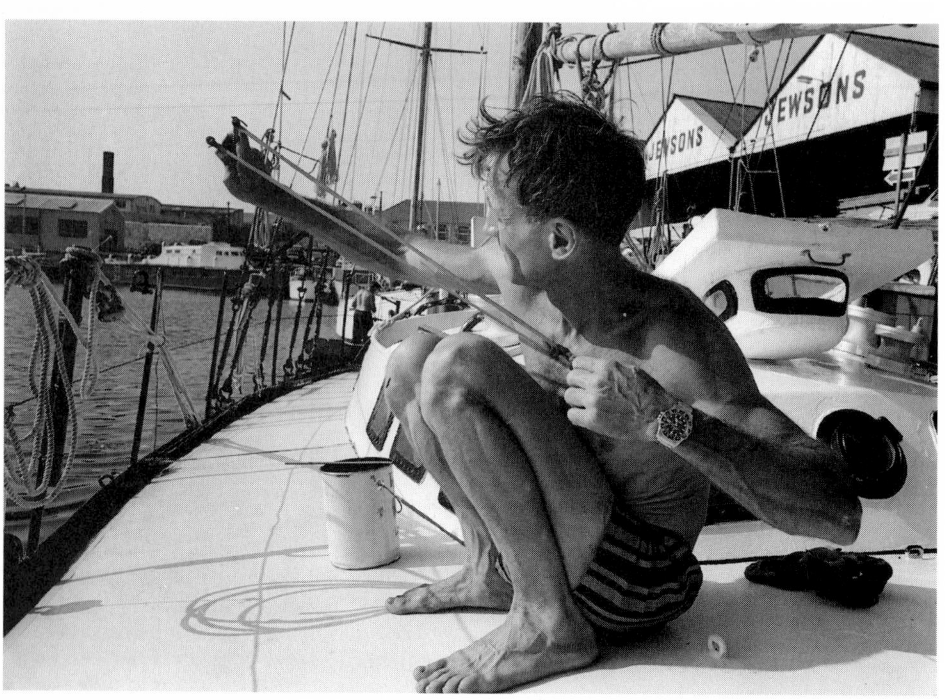

Moitessier demonstrates his chosen form of
communication at sea.

Bill King's *Galway Blazer II* was conceived for such a voyage, but not a race.

Naval officer Nigel Tetley.
He read about the race in the Sunday papers.

Tetley's *Victress.*

"I can think of nothing that was right about that boat for that race."

Loïck Fougeron under way. After his first storm at sea, the decision to drop out was an easy one.

Donald Crowhurst. "I feel like somebody who's been given a tremendous opportunity to impart a message."

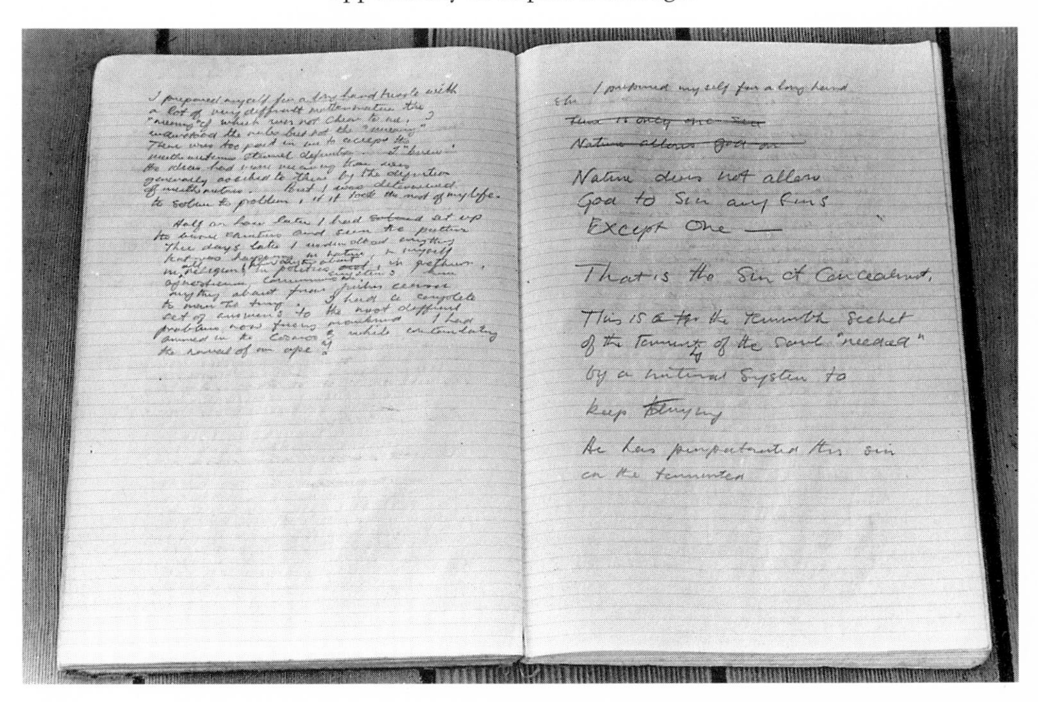

Donald Crowhurst's logbook.

Moitessier. Not since Captain Nemo had a man felt so comfortable and self-sufficient at sea.

Nigel Tetley's lonely Christmas dinner.

Crowhurst at the start of his voyage—his lines were tangled and he was towed back ashore.

Teignmouth Electron being unloaded from the *Picardy* at Santo Domingo.

The interior of *Teignmouth Electron*
at the end of its voyage.

Final resting place of *Teignmouth Electron*. Cayman Brac, 1999.

Suhaili enshrined at the National Maritime Museum. Greenwich, England, 2000.

And still farther to the east, more than 4,000 miles ahead of Moitessier, Robin Knox-Johnston was halfway across the Great Australian Bight between Cape Leeuwin and Melbourne. October 31 was a quiet day for him, and unusually warm. He had woken smelling land, and through the day saw a great number of insects and many butterflies in the air. He immediately worried that his navigation might be off and that he was far closer to shore than he suspected. But he traced the land aroma to the weed growing at *Suhaili*'s waterline that had dried and become smelly in the warmer weather, and he reminded himself that Charles Darwin had found spiders clinging to HMS *Beagle*'s rigging when the ship was hundreds of miles off the South American coast. Darwin concluded that they had been blown out to sea.

The good weather was a welcome respite from a brutal passage across the Indian Ocean. *Suhaili* had been knocked down again. The gooseneck on the main boom had broken, as Knox-Johnston had feared it would; his spare trim tab for the wind vane had broken, as the first had, and most of its parts had sunk, precluding a repair. On October 13, a southerly storm blowing great waves up from the Antarctic Ocean—"by far the worst weather I had ever encountered"—began pounding the boat so hard that Knox-Johnston thought she was going to break up. His mind raced ahead, imagining the coming disaster: *Suhaili* being smashed by the wave that would finally split her open, the cold inrushing sea, the frantic effort to break out the life raft and abandon ship, the tins of dried fruit he would grab—make sure to grab the can opener. . . . But as he was thinking this, bracing words from a Robert Service poem, *The Quitter*, filled his mind.

> When you're lost in the Wild and you're scared as
>   a child,
> And Death looks you bang in the eye,

And you're sore as a boil it's according to Hoyle
To cock your revolver . . . and die.
But the Code of a Man says: "Fight all you can,"
And self-dissolution is barred.
In hunger and woe, oh, it's easy to blow . . .
It's the hell-served-for-breakfast that's hard.

The boat wasn't coming apart yet, but he was. He was ashamed of himself. He went up on deck and watched the seas, and then took action. He streamed a large diameter polypropylene warp astern, sheeted his jib flat amidships, and *Suhaili* suddenly lay quietly with her pointed stern to the overtaking seas. No longer was she being battered, and Knox-Johnston was amazed at the difference, both in *Suhaili*'s apparent situation and his own outlook.

This was exactly the technique Moitessier had employed—and then abandoned—in the South Pacific in favor of his surfing run ahead of the waves. It didn't work for *Joshua,* Moitessier had felt, but it worked now for *Suhaili,* underscoring the truth that there is no one right way of handling storms at sea. There is only what works for different boats and their captains in different storms, an improvised alchemy of conditions and intuition.

Suhaili rode out that storm. Later, Knox-Johnston ingeniously repaired the original trim tab and once more went overboard into frigid tossing seas to fit it into place behind the rudder. He knew, however, that it was only a matter of time before it broke again, and he would be left without his self-steering gear. Then what? He would worry about that when it happened. Next he repaired the gooseneck, spending a day breaking and nursing drill bits while hand-drilling through metal plates inside the violently lurching boat. But he and *Suhaili* were both showing the wear of the voyage, and he wondered how long it would be possible to continue. Sails were splitting regularly and he was spending hours sewing with cut and calloused hands. His body was bruised, he was sleep-deprived, and he remained anxious. Australia was nearby and the land and its treats were again pulling strongly at him. He had done well in his small, rough boat, and no one would think poorly of him if he headed for port.

But he was now halfway around the world—farther, in fact, because with prevailing westerly winds from astern, his fastest route home would be to keep going. He had a commanding lead. He decided to continue as long as he could make progress. His sea heroes were watching.

Knox-Johnston, Moitessier, King, Fougeron, Tetley, Crowhurst, and Carozzo on his mooring. This was the arrangement of the race fleet of seven boats across 15,000 sea miles on October 31. Blyth and Ridgway were out. By nightfall, two more sailors would join them.

16

HOVE TO IN THE STORM during the moonlit early hours of that October 31, Loïck Fougeron curled up in his bunk, unable to sleep, waiting for what he feared would happen. He felt like a nut about to be crushed beneath an elephant's foot.

Suddenly the boat was slammed sideways by a tremendous force. The cabin's kerosene lamp went out. Everything movable—pots, plates, glasses, food, a crate of wine, books, tools, and Fougeron himself—was hurled across the cabin, which had turned on its side. In that moment, Fougeron believed he was about to die and join the multitudes of seamen who had perished far from their loved ones. He thought of his family and friends, certain he would never see them again.

But—miraculously, it seemed to him—*Captain Browne* rolled back upright. Fougeron sat on the cabin floor with blood running down his face. Going up on deck, he found the mast intact, although parts of the rigging hung slack and loose. With a surge of relief, he realized that both he and the boat had survived the knockdown intact. His next decision came to him instantly, with wonderful life-affirming clarity: he was giving up the race and heading for Cape Town.

For 24 hours, Bill King had stood in his (belowdecks, protected) cockpit, watching this same eerily moonlit storm through *Galway Blazer*'s two round Perspex hatches. For a while, the wind vane held the boat on its downwind course, but as the wind and seas rose it was overpowered and King had to steer by hand. As the storm progressed, he had reefed the junk-rigged sails—he was able to do this with lines led belowdecks—until he was finally running downwind under bare poles. It was the most violent weather he had ever experienced. He estimated the waves at 40 feet in height. At the top of one wave he looked down into the trough dropping away before the boat and saw a petrel flying across a patch of moonlight far below him.

At 9.30 A.M. on October 31, the storm reached its furious height as the wind backed to the west. With the sudden change of wind direction, the waves lost their regular pattern in a confusion of tumbling cross seas. At that point, King stopped running and allowed the boat to lie ahull—that is, with its beam to the seas, the position most boats will assume when left unattended with no sail up. King felt *Galway Blazer* rode as well like this as she had when running downwind. By the evening of October 31, the wind began to die down, and King believed the worst was over.

He decided to go out on deck to try to find out why the wind vane had been locking. Conditions had moderated sufficiently for him to leave both plastic cockpit hatches open. He went aft, but could see nothing wrong. However, he noticed that the foresail needed lashing, so he went below again to get a length of rope.

> I was sitting jambed [sic] into place under the open hatches, coiling down the rope, when . . . over she went to 90 degrees. The boat was now lying right over on her side. Hurled by the elemental forces of the breaking peak of a rogue sea mountain, she was using her side as a surfer would his board. The masts

> must still have been in the air, in their proper element, and I had time to think, "She will come back again; that great lead keel will swing her upright."
>
> Even as the thought crossed my mind, a vast new force started to act upon us. In those confused seas there was no proper pattern. Some cross-riding protuberance of foam-lashed water rode across the trough in which we might have recovered. Into this obstacle our mast tops now buried themselves, driven by the frightful impetus of our sideways rush. The leverage of a new element, imposed on our mastheads, now started the action of the mariner's most dreaded catastrophe: a complete rollover, upside down.
>
> I had a rapid change of mind. "She will come back again" became "No, she won't"; and, indeed, she did not.
>
> I was on my shoulders pressed against the deck head, which was normally above me, my head pointing to the sea bottom, fifteen thousand feet below, looking at green water pouring up through both hatches.

Poised upside down, with eternity beckoning, time elongated like a rubber band—and then snapped right back again: the 2-ton keel wrenched *Galway Blazer* back upright.

Water was calf-deep inside the boat. King pulled the hatches closed and began pumping out the water. When the pump sucked air in the bilges, he went on deck again, and saw the extent of his disaster. The foremast had broken off about 12 feet above the deck. The mainmast was still intact, though fractured, and pulled over to starboard at an angle. The wind vane was smashed.

Bill King's voyage was equally shipwrecked. *Galway Blazer* could never sail through the Southern Ocean as she was. The best he could do now was to limp toward Cape Town. It was a crushing end to a dream.

But King knew he'd been lucky: if the capsize had occurred 60 seconds earlier or later, he would have been on deck, washed overboard, or smashed in the wreckage of his foremast.

It is unfair to compare the outcome of two very different boats handled by two skippers in the same storm. Every moment at every geographic spot at sea is a unique combination of forces; add two boats several hundred miles apart and the varying factors become infinite.

Nevertheless, the management of small boats in heavy weather is of paramount importance to sailors, and it is irresistible to review the results of the different tactics employed by Fougeron and King in their dissimilar boats in the same weather event.

It seems clear that even at the height of the storm, Fougeron's boat, *Captain Browne,* still carried reduced storm sails. "How long can the sails hold out in this fury?" he wrote in his logbook. The wind vane was disconnected, and Fougeron was below in his bunk, not steering, not trying to make progress, but riding out the storm with the sails arranged so that *Captain Browne* was hove to.

In similar conditions, King allowed *Galway Blazer* to lie ahull, that is, with no sail up, beam-on to the seas.

Heaving to is a time-honored method of stopping a sailing vessel in the water when wind would otherwise push it on. Pilot vessels once hove to on station while waiting for ships outside of ports and harbors. Small yachts overtaken by gales can heave to with remarkable comfort until conditions improve. This is simply done by arranging sails and helm in such a combination that the boat stalls comfortably and sustainably, with its bow pointing about 50 degrees from the eye of the wind, making perhaps half a knot of drift at right angles to the wind. In this position a boat does not present its entire broadside, its most vulnerable aspect, to the oncoming waves, but points obliquely into them, parting them—even large breaking waves—and riding over them. Small sailboats are easily hove to by setting a reefed portion of the mainsail, or a storm trysail (a small ruggedly made sail dedicated to storm use set in place of the mainsail), together with a small headsail backed to windward: the small main or trysail tries to drive the boat forward and into the wind; the backed headsail opposes this drive, preventing the boat heading to windward and effectively stalling its motion. The helm (tiller or wheel) is used to fine-tune the boat's balance, augmenting the force of either the main or trysail or the headsail. This may sound complicated, but it's not—though it requires practice. It is often said that modern sailboats with their fin keels cannot heave to properly, but this is not so. Every boat behaves slightly differently, and its captain or crew must learn its qualities and experiment with its heaving to rig by tweaking sails and helm.

In a gale, the difference between a boat running downwind at high speed (corkscrewing wildly over the waves) or beating hard to windward (constant bone-jarring pounding) and a boat hove to in the same conditions has to be experienced to be believed. With the speed and violent motion of progress come the fear and anxiety for one's own safety and the boat's structure. Hove to, noise and motion are amazingly reduced, and hot drinks and meal-making can be possible where minutes before they were unthinkable.

A further, almost magical component of heaving to, is the slick, or wake left by a boat's hull. This is the area of sea immediately to windward, between the boat and the oncoming waves, created by the wind pushing the stalled, resisting boat slowly through the water. The water surface in the boat's wake appears slightly disturbed, like the water on one side of a moored buoy in a strong tidal stream, and almost glassy, like an oil slick. This has the astonishing effect of interrupting the heaping waves as they reach it. Large breaking seas are suddenly tripped by the slick, lose their height and power, and tumble harmlessly before reaching the drifting boat.

The wonderful skill of heaving to was once part of the seaman's standard bag of tricks, learned from older salts or on the decks of ships. It was well-understood and routinely employed. But with modern yachts being purchased as easily as cars and no license required to sail them, this vital piece of seamanship is often forgotten about, or not learned, or learned improperly. It's also understandable that few recreational sailors will take the time to head offshore into bad weather for the purpose of trying out the heavy-weather tactics they've read about in the books.

Yachts are marvels of engineering; their three dimensions of compound curves are arranged in beautiful sculptural shapes designed to accommodate the sea in so many of its moods. They generally behave so well that it's easy for the inexperienced sailor to believe that a good sea boat will fend for herself in large seas. And in conditions where wind has risen past a point where fully reefed storm sails still make a boat feel overcanvased, it can seem reasonable and prudent to remove sail altogether. The boat will then lie beam-on to the seas, and—in most cases—this will result in no more than discomfort. Few recreational sailors will ever experience conditions that will demand a true seaman's reservoir of knowledge and skill in order to survive. Thus, the easy and usually adequate "technique" of lying ahull becomes the favored storm tactic of most small-boat sailors.

Yet in severe conditions, lying ahull can be deadly. The fourth edition of *Heavy Weather Sailing,* the classic book on storm tactics at sea by the late English author and sailor Adlard Coles, and Peter Bruce, clearly presents the conditions—proven through tank testing—under which lying ahull will result in disaster.

> It is breaking waves that cause capsize. If the yacht is caught beam-on to breaking waves of sufficient size . . . a full 360 degree roll will be executed. How big do breaking waves need to be to cause this type of behaviour? Unfortunately, the answer is, not very big. During the model tests . . . when breaking waves were 30% of hull length, from trough to crest, they could capsize some of the yachts, while waves to a height of 60% of the hull length would comfortably overwhelm all of the boats we tested. In real terms this means that for a 10 meter (32 ft 10 ins) boat . . . [a] breaking wave 3 meters (9 ft 10 ins) high . . . presents a capsize risk, and when the breaking wave is 6 meters (19 ft 8 ins) high, this appears to be a capsize certainty for any shape of boat.

The book goes on to report that the same model boat, differently aligned so that it is not beam-on to the breaking wave (but, rather, pointing obliquely into it, for instance, as if hove to) will not capsize.

Heavy Weather Sailing is filled with stories of boats lying ahull being rolled over, capsized, dismasted, of crews being lost. But it contains only one such incident while a boat was hove to. In the disastrous Fastnet Race of 1979, which was disrupted by a strong gale, 158 boats out of a fleet of 300 adopted storm tactics: 86 lay ahull, 46 ran before the wind either under bare poles (Moitessier's choice) or towing warps (Knox-Johnston's tactic), and 26 hove to. One hundred of these boats suffered knockdowns, 77 were rolled over at least once. Not one of the boats that hove to were rolled, capsized, or reported any major damage.

Bill King, the navy commander who had cruised the world in submarines and sailed across the Atlantic, chose to lie ahull during his storm, and *Galway Blazer* was capsized and rolled over, wrecking its rig.

Loïck Fougeron, who didn't have the years at sea under his belt that King had, adopted the tactic of heaving to in his smaller, heavier boat, which was knocked on its side, but otherwise suffered no major damage.

It is just possible that tactics made the difference.

When their great storm abated, Fougeron steered for Cape Town. But headwinds and cold finally dissuaded him and he turned north, making his first landfall in three months on November 27 at the southern Atlantic island of Saint Helena. There he found much kindness from the local doctor and residents.

Bill King raised the two struts that had been handily built to lie on *Galway Blazer*'s deck in case of the loss of his unstayed masts, and headed for Cape Town under jury-rig. He was in daily radio contact with friends in England and Cape Town and sent messages to his family, keeping up a cheerful front. But he was deeply depressed by what had happened and confided this to the log he was writing to his wife, knowing she would only read it long afterward.

November 8th, Friday

My Darling,

I sent you a message [via radio] *about how well I was taking the disappointment. Then I was probably in a state of euphoria, after having my life spared by thirty seconds. I do not think that survival would have been possible had I been out on deck tending the foresail when we turned over.*

As the danger recedes, I get broody. I realize the cold facts. My voyage has been stopped. My little boat lies broken, and I am alone with my bitter disappointment, creeping along at perhaps fifty miles a day. I knew such an adventure must be dicey, but I never gauged how shattering a blow this disaster could deal my spirit.

He reached Cape Town on November 22.

S

17

DONALD CROWHURST RECORDED in his logbook that he was seasick all through his first night at sea and most of the following day, as he tacked west down the English Channel toward the Atlantic. He put it down to nerves.

His first job was to store the jumbled mass of loose gear, equipment, and food that lay in heaps on his bunk, on the cabin sole, on his table, everywhere throughout the boat. One of Crowhurst's few successes with a sponsor had been the Tupperware company. Now he filled scores of plastic containers at random with whatever came to hand—food, tools, batteries, film, hardware—and stacked them on shelves on either side of the boat's one single bunk, forward of the saloon. Beneath the bunk and the saloon seats and into every pocket of the main hull he stored more food, life jackets, flares and signal flags, the film camera given to him by the BBC, sailing manuals, instruction manuals, water jugs, his harmonica, sextant, medical supplies, hot water bottle, pilot books, and the few books he had brought with him: technical reads with titles such as *Servo-Mechanisms, Mathematics of Engineering Systems* and a couple of sea books, *Shanties from the Seven Seas* and Chichester's Gypsy Moth IV *Circles the World.* He had told Clare he didn't want any novels. Instead, for inspirational reading, he had brought *Relativity, the Special and General Theory,* by Albert Einstein.

He had enough Tupperware containers filled with electronic parts to start a small factory: boxes and boxes of transistors, condensers, resisters, switches, valves, circuit boards, wire, plugs, and sockets. These were things over which Crowhurst had mastery; they were the components and currency of his particular brilliance, out of which he could always fashion order no matter what chaos lay scattered elsewhere. The abundance of such supplies must have been a comfort to him.

Everywhere inside *Teignmouth Electron*'s cabin ran neat streams of color-coded wire, fastened to bulkheads, the cabin top, running between the masts and the hulls, all according to the complicated wiring diagrams Crowhurst had provided Eastwoods for the working of his computer-operated electronic process control systems. All these wires came together in a thick confluence that ran down the port side of the cabin and disappeared under a red seat cushion. Beneath the cushion the wires ended in an unconnected tangle—in the empty space where Crowhurst's computer was supposed to have been. In the scramble to get his boat built and to depart in time, his computer—his "box of tricks" he called the revolutionary device that would sense the boat's condition, adjust the sails, set off the buoyancy bag, enable Crowhurst to sail his trimaran at breakneck speeds and make the fortune of Electron Utilisation—had not been produced. It was still just a dream.

On his third day at sea, despite the early-in-the-voyage overabundance of stores, he was already worried about his supply of methylated spirits, used to prime the burners of his kerosene stove. He had calculated quantities from figures given for the needs of two people in Eric Hiscock's *vade mecum* reference book *Voyaging Under Sail,* much of which was based on Hiscock's world-girdling voyages made with his wife, Susan. Crowhurst had systematically halved the amounts suggested, forgetting that one voyager will use a stove as often as two.

Still, he calculated, and wrote down, that he had enough to last 243 days. He need not have worried: his voyage would last exactly 243 days.

As he packed things away and tried to arrange his stores over a period of several days, he also had to navigate and handle sails and keep the boat moving, and he discovered, in quick succession, a cascade of important failures. His Blondie Hasler-designed steering gear was an early and consistent problem: screws and bolts from it began to work loose and disappear. The gear had been quickly and poorly installed—as had been the lifting genoa track on deck, noticed days earlier by the BBC cameraman. Electrical parts he had aplenty, but Crowhurst had brought along virtually no spare screws and bolts, so he was forced to take screws from other places on board to use on the wind vane gear. Those too soon worked loose and disappeared overboard, infuriating him. "That's four [screws] gone now," he wrote. "Can't keep cannibalising from other spots forever! The thing will soon fall to bits!" Then he cut a finger on his left hand while trying to hoist a metal radar detector. "Blood everywhere—first aid kit out. Certainly well stocked in this department!!"

On Tuesday, November 5, he noticed bubbles blowing out of the hatch on the bow of the port hull. He opened it up to find the bow compartment flooded to deck level with water—a galvanizing sight. He quickly bailed the water out with a bucket. The problem, he hoped, was not the hull but the seal of the hatch, and he screwed its wing nuts down again over a new fiberglass gasket.

Less importantly, but far more demoralizing for him, he was having trouble with his radio equipment. He could not pick up signals on his Racal receiver and spent hours taking it apart. Then he couldn't raise Portishead Radio on his Marconi transmitter.

He made painfully slow progress out into the Atlantic. From November 2 to 6 he sailed 538 miles according to the readings on his log, which was a fast 134.5 miles per day, suggesting good progress. However this was mileage covered while tacking south and west, and the true distance made good along his route was 290 miles—an average of 72.5 miles per day.

Despite the troubles aboard, Crowhurst never forgot that he had embarked on an extraordinary voyage, something well beyond the scope of what most people might ever experience. BBC Bristol had given him £250 and a 16-millimeter camera, film, and a tape recorder to make a film of his voyage, and though he did little filming to begin with, he soon began making tape recordings. Crowhurst took seriously the charge to bring home a record, and he looked well beyond a simple description of his daily routine. Sailing around the world alone in a small boat would, he believed, prove to be a seminal experience, and he wanted to make sure he got it down on film and tape. "I feel like somebody who's been given a tremendous opportunity to impart a message," he recorded soon after leaving England, "some profound observation that will save the world."

On November 13, Crowhurst found that the "waterproof" hatch on the cockpit floor, which had leaked on the maiden voyage but had supposedly been repaired by the Eastwoods crew at Teignmouth, was again leaking badly. He had been pushing south against strong head winds and water had streamed aboard, repeatedly filling the cockpit, which although fitted with drains, drained slowly. Seawater had flooded the engine compartment immediately below the hatch, soaking his generator and the bulk of his working electrics. For Crowhurst, this was a disaster graver than a leaking hull. The possibility of being unable to produce electricity completely undermined him.

The growing reality of his adventure, which he had so forcefully, cleverly brought on himself, risking everything—bankruptcy, the well-being of his family, his self-respect, and his life—crept upon him in the cold, wet cabin where he now found himself alone, somewhere at sea off the wintry coast of northern Europe, with devastating starkness. His boat had begun to fall apart even before he left port and had been breaking down ever since in reasonable, if unpleasant weather. The prospect of Cape Horn and the Roaring Forties was now a grim one.

Crowhurst's reaction was commendably sane. He considered, perhaps for the first time, giving up.

> *Friday 15 (November)*
>
> Racked by the growing awareness that I must soon decide whether or not I can go on in the face of the actual situation. What a bloody awful decision—to chuck it in at this stage—what a bloody awful decision! But if I go on I am doing [two] things:
>
> 1. I am breaking my promise to Clare that I would only continue if I was happy that everything was as it should be to ensure the safe conclusion of the project. Unless I can get the electrics sorted out, I cannot honestly say the conditon is met. Furthermore I am placing Clare in the horrible position of having no news of me for 7 to 9 months, as the radio would not be functioning.
>
> 2. As the boat stands, I cannot drive her much above 4 knots in the 40s. The Hasler performs wild broaches that would be fatal in big—really big—seas, when running. . . . I cannot reasonably see a fast passage in the 40s in safety without self-righting gear, the buoyance bag device, in operation. Particularly bearing in mind that I started late . . . as it means arriving at the Horn far later than I anticipated, in six or seven months' time—April/May [nearing the southern winter]. With the boat in its present state my chances of survival would not, I think, be better than 50–50, which I would not regard as acceptable. "The boat in its present state"—what does that mean?

He listed the problems. First was the possibility of not being able to generate electricity. If he had no electricity he would have no radio communication, no masthead buoyancy, no time signals, no light.

Leaky hatches had let in 120 gallons in 5 days. The cockpit hatch had leaked 75 gallons overnight. The only proper solution—screwing it down permanently—would seal off the generator and shut down the boat's electrical system.

Far worse than this, he had no way of pumping out the leaking hulls. Getting water out of the inside of a boat, where, it will inevitably find its way sooner or later, is a fundamental principle of seaworthiness. But Eastwoods had not installed the hose for his bilge pump, which rendered it useless. The only way Crowhurst could get rid of the large quantity of water leaking into the boat's three hulls was to bail it out with a bucket. This could hardly be done in bad weather, when water would be most likely to come in.

It was a thorough, rational list, chilling in its presentation of the situation. Inside his logbook, he wrote pages of arguments for and against several options: he could return to England and try again the next year, at least for a faster time—except that Stanley Best had already paid out far more than he'd ever expected to and would be unlikely to support the project through another year. Losing Best's support meant more than just an end to funding: Crowhurst's business and house were now virtually owned by Stanley Best, who, Crowhurst feared, had every reason to call in the debt. Another idea was to save face, and perhaps boost the value and notoriety of *Teignmouth Electron,* by sailing it as far as Cape Town or Australia and selling it there. But this seemed a slim possible benefit at the end of a long, hard voyage.

Crowhurst argued back and forth with himself on paper, but all ideas, all possible alternatives, resulted in the same unacceptable conundrum: returning home would result in shame and bankruptcy, yet to sail on appeared profitless and dangerous. He could not bring himself to make a decision.

> I will continue south and try to get the generator working so that I can talk to Mr. Best before committing myself to any particular course or retiring from the race. I suppose I'm just putting off the decision? No. It's far better that he should know before I commit the project to withdrawal, and that I should have his views. If he doesn't want anything further to do with

the nonstop project (as distinct from the S. T. race) things would be really black—but at least I'd know where he stood. In the final analysis, if the whole thing goes quite sour: Electron Utilisation bankrupt and Woodlands sold, ten years of work and worry down the drain, I would have Clare and the children still and:

If you can make a heap of all your worries
And risk it on one turn of pitch and toss
And lose, and start again at your beginnings
And never breathe a word about your loss

18

ON NOVEMBER 3, 300 MILES southwest of Melbourne, Australia, *Suhaili*'s self-steering trim tab broke again; the metal shaft sheared off and the bottom of it sank into the ocean. This was the original trim tab, which Robin Knox-Johnston had repaired and shipped back into place when his spare had similarly broken and sunk. This meant the end of his self-steering gear.

Suhaili had received a terrible battering in the Indian Ocean. The constant flood of seawater into the boat from the hatches and around the cabin edges had taken its corrosive toll below, knocking out his Marconi transmitter six weeks before. Rust stains from the rigging streaked the hull. Important parts of the boat were now held together literally with string. The tiller had snapped off, and the rudder was loose on its pintles (the hinge-like bearings on which the rudder hung), so Knox-Johnston had lashed the spare tiller to the rudder head and wound more line between the rudder and the metal pushpit tubing around the stern to hold it in place in case the pintles broke. Had she been in the English Channel, any sailor coming upon *Suhaili* looking as she did now would have presumed her to be a vessel in distress and offered to tow her into the nearest port. Knox-Johnston's round-the-world effort gave every indication of being pulverized into submission, and with the long Pacific leg of the Southern Ocean ahead and Cape Horn at the end of it, heading for Melbourne made compelling sense.

But he was in the lead. Despite her appearance, *Suhaili*'s "icebreaker" hull was intact, her masts and rigging stood fine, and Knox-Johnston was bruised but healthy. He decided to see if he could press on and get as far as New Zealand.

It was the lack of self-steering, the idea that he might have to sit in the cockpit and steer for at least 16 hours a day, that most daunted him. By then, including the voyage from India, he had sailed about 32,000 miles aboard *Suhaili,* and it was all the sailing he had done. Knox-Johnston was not a yachtsman, but a seaman, a master mariner, who had chosen, for pragmatic rather than recreational reasons, to go voyaging in a small wooden boat. In this he was no different from Joshua Slocum, who had returned to the sea in his *Spray* in 1895, and had sailed that fat, unhandy craft around the world with no self-steering gear. Slocum had discovered, by experimenting with his sails and helm, that *Spray* could be balanced to hold a course on any point of sail in most conditions—or so he had claimed, to the amazement and frequent skepticism of yachting pundits ever since. Knox-Johnston remembered this and took heart. *Suhaili,* like many long-keeled boats, balanced nicely when sailing to windward and would hold such a course for hours without a hand on her helm. It is with the wind on the beam and from astern that boats show a tendency to slew off course and come around into the wind.

When the trim tab broke, the wind was northerly, and Knox-Johnston wanted to head east. He lashed the helm amidships, neutralizing it, and then began to play with the sails. *Suhaili,* a ketch with two masts, had a basic wardrobe of four sails: main, mizzen, and two headsails, the staysail and jib. In addition, Knox-Johnston carried six more headsails of varying sizes, from light-air spinnaker to tiny storm jib, which he could add, substitute, or subtract as conditions warranted. With the long bowsprit, this meant that the boat's propulsive force was spread out across and beyond its entire length, giving Knox-Johnston a wide patchwork field of canvas to tweak and play with in his search for a balanced trim.

But for the next few days the wind offered no challenge. It remained northerly and northeasterly and *Suhaili* easily held her course to windward, giving Knox-Johnston plenty of time below for sleep. Not until November 6 did the wind finally come from astern, southwest, and by then he was approaching land. He was closing with Bass Strait, the 50-mile gap between King Island and Cape Otway on the southeast coast of Australia.

Or he hoped he was. It had been four months since he had last glimpsed land when passing the Cape Verde Islands in the North Atlantic—four months in a vast isolation tank of wind and water devoid of all hint of the world beyond the constant encircling horizon. Since then his sense of where he was on the face of the globe had been entirely derived from his navigation—a mathematical supposition clung to in the absence of any empirical proof. His sextant had been soaked by sea spray and its mirrors had become tarnished, and he wondered now about its accuracy. As he sailed into the strait, he looked for land birds and the types of clouds generated by the heat of land masses, or any sign of the land he wanted to believe lay just over the horizon to the north and south, but he saw nothing. An absurd idea pestered him: What if he was nowhere near Australia? What if his navigation was wrong, or distorted by some cumulative error?

But then he remembered the radio stations he'd been listening to for the last few weeks: first Perth, on the west coast; then Albany, a coastal outback town serving vast sheep stations of the interior, with its news of wool sales; and now he was picking up Melbourne. He had to be where he thought he was.

At 2230 that night, he saw a flicker of light to the east. Just before midnight he identified it as Cape Otway light. He sat in the dark and steered through the rough waters stirred up around the cape, feeling intensely pleased. He had come as far

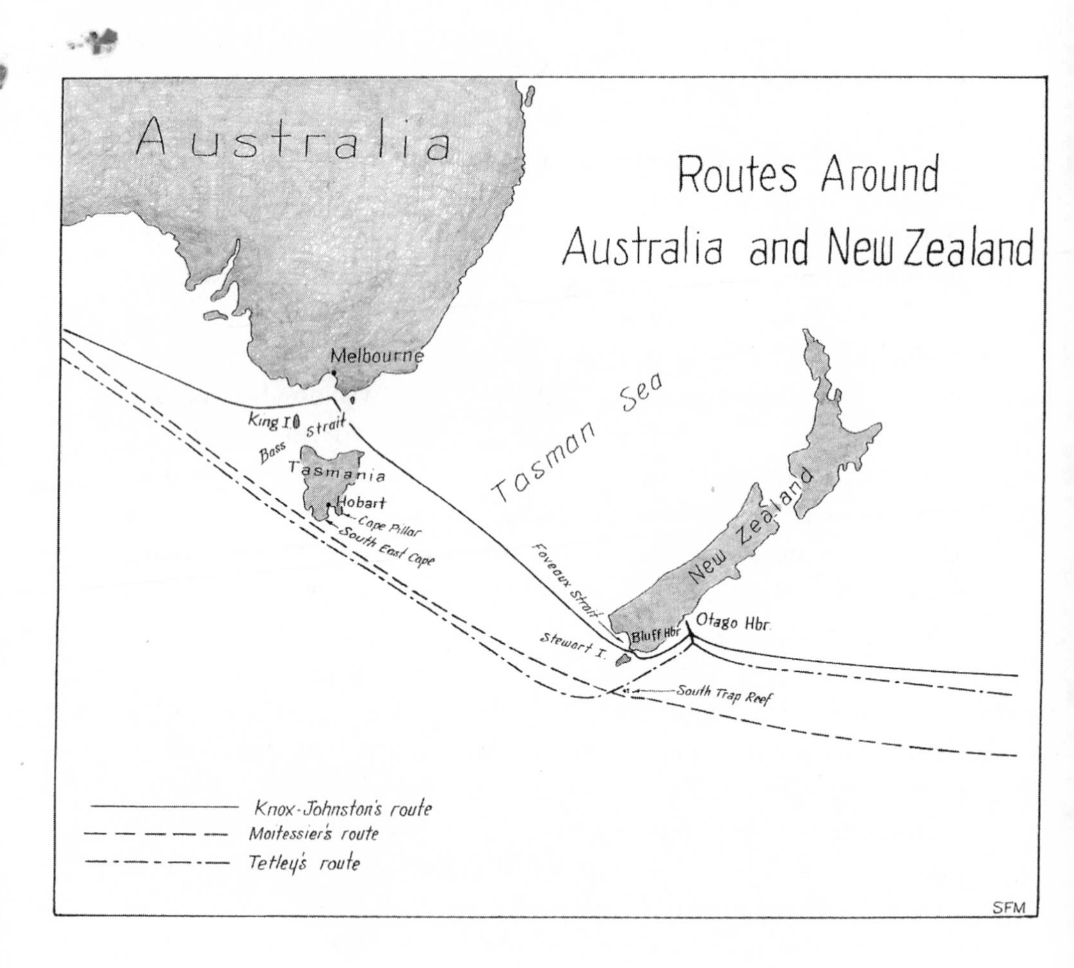
Australia
Routes Around
Australia and New Zealand
Melbourne
Tasman Sea
King I.
Bass Strait
Tasmania
Hobart
Cape Pillar
South East Cape
Foveaux Strait
New Zealand
Bluff Hbr
Otago Hbr.
Stewart I.
South Trap Reef
Knox-Johnston's route
Moitessier's route
Tetley's route
SFM

as Chichester without stopping, and in a much smaller boat. He thought of that half-forgotten world ashore, where people washed regularly, ate and slept well, and kept company with each other. But now pulling in offered no real temptation as long as he could keep going.

Once clear of Cape Otway, he got *Suhaili* sailing herself before the wind (the most difficult point on which to balance a boat), with her booms sheeted wide like wings, and turned in for three hours' sleep.

In the morning, the weather was fine and Australia stood clear above the sea to the northwest. By early afternoon, Knox-Johnston was closing with Port Phillip Heads, the entrance to Port Phillip Bay and Melbourne harbor. He was hoping to find a boat that would forward his mail on to London and let his family, friends, and sponsor know that after almost two months of radio silence, he was well. Soon enough he spotted a pilot vessel approaching an inbound ship. He was able to get close enough to the pilot to shout that he was sailing nonstop from England and tossed over a waterproof box full of letters, charts, film, and articles written for the *Sunday Mirror.* Also in the box was a message to the *Mirror* suggesting the small port of Bluff, on New Zealand's South Island, as a possible rendezvous for the next mail exchange.

By the next morning, he was off the north coast of Tasmania, heading for Banks Strait, a 7-mile-wide shortcut through the islands at the eastern end of Bass Strait that allowed him to head more directly to New Zealand. It was a clear sunny day. The great green heart-shaped bulk of Tasmania now lay between him and the ever-present southwesterly swell of the Southern Ocean, and for the first time in two months the sea was reasonably smooth. Knox-Johnston took this opportunity to pull himself up to the top of the mainmast in his bosun's chair to collect the slides that had come off the mainsail and remained jammed at the top of the track. While he was doing this, a small plane appeared out of the blue sky. It swooped down and circled him for half an hour. He hoped this meant that his mail and messages had been passed on, and that maybe the plane had been chartered by the *Sunday Mirror.* Slides retrieved, he sailed slowly on.

Later, while he was sunbathing nude in the unaccustomed sunny weather, the aircraft returned, circling *Suhaili* again, and during the afternoon, another airplane and a helicopter appeared, circling and taking a close look at him. He was news, he realized. Sure enough, after his long silence, the position and photographs of the race leader appeared in London's Sunday newspapers on November 10.

Thirty-six hours later, a little after midnight, Knox-Johnston passed the winking light of Swan Island at the eastern end of Banks Strait. He put *Suhaili* on a course across the Tasman Sea toward Bluff, on Foveaux Strait, South Island, New Zealand.

For the first two days in the Tasman Sea the wind was easterly, and *Suhaili* happily steered herself close-hauled. But then the wind moved into the west again, its prevailing quarter, and Knox-Johnston set about learning in earnest how to sail his boat without self-steering gear. Over the next few days, experimenting with every combination of sail, reefed and unreefed, he discovered as much again about *Suhaili*'s sailing abilities as he had learned in all the years and tens of thousands of miles in which he had previously sailed her. He found that she could run and reach off the wind—as Slocum's *Spray* had been able to do—for long periods under reduced and balanced canvas, long enough for him to get sufficient sleep before she gybed and threw him out of his bunk below. The boat had always possessed these abilities, but it had required necessity and the abandonment of other methods to discover them. This is what sailors have always done as long as they have gone to sea in boats, and it is only the recent invention of efficient self-steering systems that has brought about the widespread atrophy of this skill in modern sailors.

Suhaili's daily runs across the Tasman Sea were, in fact, as good as they had been before the self-steering gear had broken. Knox-Johnston was getting enough sleep—12 hours straight one night—and now his hopes rose as he began to realize the voyage could go on. The Pacific and the Horn no longer looked so impossible.

On Sunday evening, November 17, Knox-Johnston was listening to the weather forecast from a New Zealand radio station. New Zealand's weather comes to it across the Tasman Sea, and great attention is paid to it on radio broadcasts, giving Knox-Johnston the most accurate and up-to-date weather information he had had on the voyage so far. The report that night noted a deep depression forming south of Tasmania. Its inevitable direction would be across the Tasman Sea, resulting in a gale within the next few days or sooner. Knox-Johnston hoped it would pass well to the south, taking the path of a previous forecast low. The thought of closing with land during a gale worried him.

Usually, he switched the radio off as soon as the weather report was finished, but that evening, busy with something in the cabin, he left the radio on for a few minutes, and was startled to hear the following words: "Master, *Suhaili* . . . Imperative we rendezvous outside Bluff Harbour in daylight. Signature: Bruce Maxwell."

Knox-Johnston was thrilled. His messages had got through. Bruce Maxwell was the *Sunday Mirror* reporter who had sailed with him aboard *Suhaili* from London to Falmouth, a founding member of the *Suhaili* Supporters Club, and he was excited at the thought of seeing him soon and getting mail from home.

The weather was the only problem. Bluff was then about 100 miles away. With luck, he could get there before dark the next day—otherwise he would have the unhappy choice of waiting through the next night, risking being caught by the onset of heavy weather, or abandoning the chance to see Bruce and receive mail from home and scudding off to the safety of the open sea. But he believed he could beat the weather, so he held his course.

His alarm clock woke him at 0500 the next morning.

Land—the rugged outline of New Zealand's mountainous South Island—was visible far off to the north, and Solander Island, a high, uninhabited rock, stood up out of the sea about 15 miles to the the east. Solander was 72 miles from Bluff, so Knox-Johnston was immediately able to pinpoint his position.

He soon realized he wouldn't make Bluff before dark, so he hove to for most of the morning near Solander Island, getting underway again at noon. This, he thought, would get him to Bluff in daylight early the next morning, and still ahead of the approaching depression.

He was fully aware that he was heading into a dangerous bottleneck. Bluff lay at the north side of Foveaux Strait, a narrowing body of water separating South Island and Stewart Island, full of dangers for a boat. The eastern end of the strait was awash with small islands and shallows, which could throw up steep breaking seas in rough weather. To the north, near the western entrance, were more islands. It was also the fastest way past the bottom of New Zealand, avoiding a long detour around Stewart Island, and Knox-Johnston thought he could slip through Foveaux Strait, grab his mail, and be out into the Pacific Ocean in daylight, before the weather turned. This was his last chance to turn away for the open sea, leaving Stewart Island to port, and he decided to gamble. He committed himself to this course—a commitment because *Suhaili* was not a weatherly boat: She could not point close to the wind or make much headway against it when strong. She needed searoom when the wind blew against her, so once into Foveaux Strait, she could not easily sail out the way she had come in. She would have to keep going.

As he entered the strait, the barometer began to plunge. That morning, as he neared Solander Island, it read 996 millibars. By late afternoon it was at 980 millibars, and Knox-Johnston realized then that the depression had moved much faster than he had thought it would. The steep barometric fall told him that the system's isobars were close together, the depression was very deep, and the wind, when it came, would be violent. It

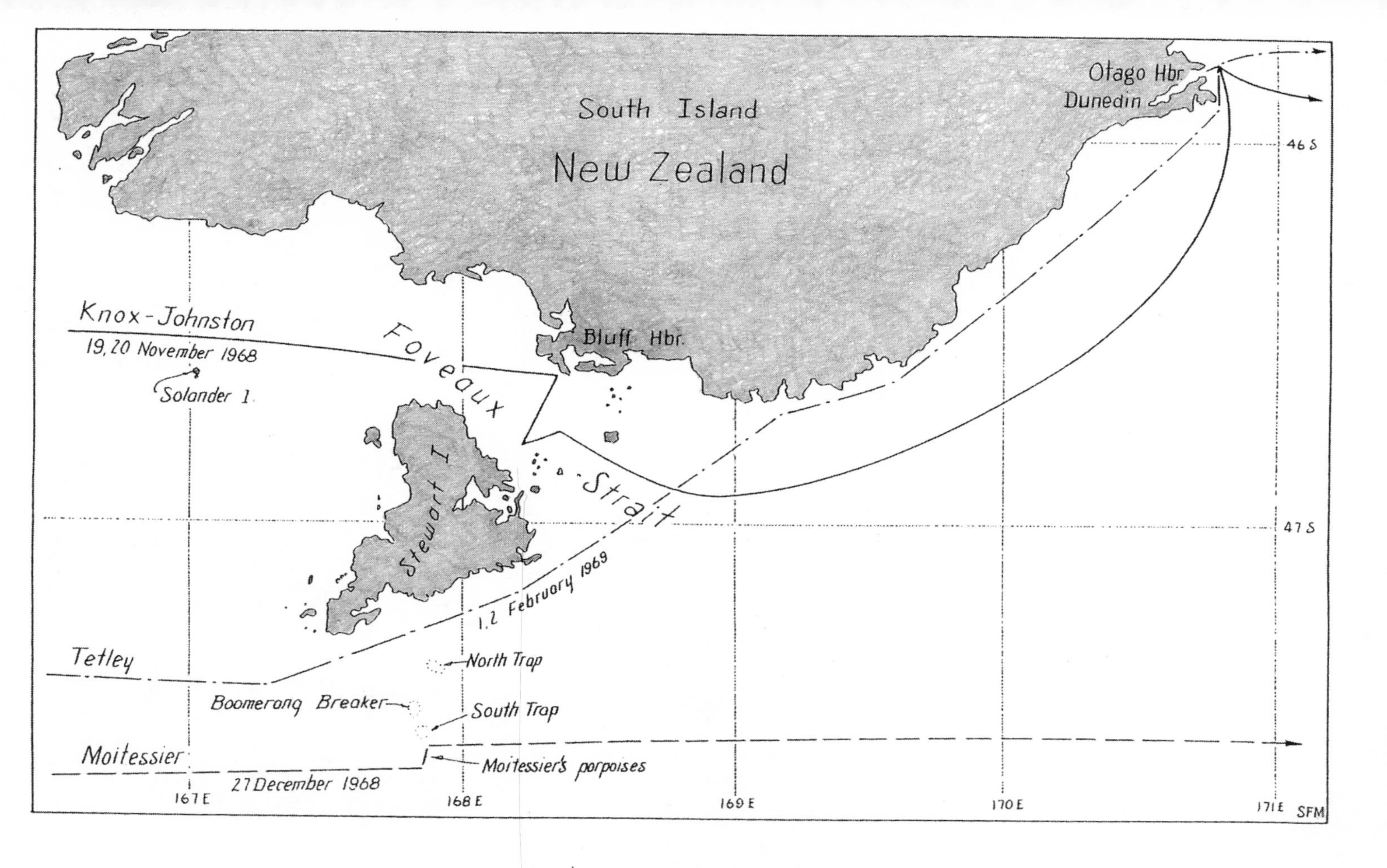

Routes Around New Zealand

would start to blow from the north, veer to the west, and then blow hardest from the southwest. This would make Bluff, and South Island, and the many small unlit islands at the narrow eastern end of Foveaux Strait, through which he was now committed to pass, all a lee shore—the shore a seaman dreads the most, because if he cannot find his way through some gap in it, he must hope to claw off it.

At dusk, which came early as dark clouds obscured the sun in the west, Knox-Johnston began to do what he could to prepare for the weather. He got out his sea anchor and heavy warps, ready to stream astern if he needed to slow down. He worried that the two now useless plywood wind vanes could cause damage in heavy seas, so he threw them overboard. He lowered the mainsail and lashed it tightly to its boom. The wind was still light, the sea quiet, and *Suhaili* sailed slowly and comfortably on into the strait under reefed mizzen and storm jib.

It was time again for the radio weather report. The announcer—no doubt in a warm studio with a cup of coffee or tea at hand—calmly read out the details of the coming storm. The system's cold front was 80 miles away, moving east at 40 knots: its leading edge would reach him in 2 hours. Winds of 40 to 50 knots were forecast, increasing to storm force of 55 knots and higher the next day. Heavy rain and poor visibility would come with the storm. Again, he heard the announcer read out the message from Bruce Maxwell, and he wished he'd never heard it because then he would have been through Foveaux Strait and into the open sea by the time the storm hit.

He thought of Maxwell at this moment probably drinking a beer in a hotel bar ashore, and Knox-Johnston wondered if he'd be drinking with him in 24 hours' time. He even began hoping he might, and then he felt ashamed of his thoughts. He went to work.

He made himself a cup of coffee, put some brandy in it, and pored over his charts, and the Admiralty pilot book and light list for the area, fixing in his mind what would soon lie around him in the dark, and what lights to look for to keep track of his position through the night—if he was lucky enough to see them in the coming storm. He had to keep two dead-reckoning plots going in his head, allowing for the greatest and least possible set of the tides in the strait.

Then he stood in the companionway hatch, half in and half out, smoking a cigarette as *Suhaili* sailed through the dropping dark. As the weather bulked and packed in astern, he saw at last a flicker of light ahead, which he identified as the light on Centre Island, 25 miles northwest of Bluff. Then the light disappeared as rain began to pelt down, and the dark closed in around him.

With the rain came a shift of wind, into the west, astern, and it began to strengthen. Knox-Johnston wanted to go as slowly as possible so he streamed his sea anchor and big blue polypropylene warp. This also helped steer the boat, keeping its stern to the waves and rising wind, while the storm jib sheeted flat on the bowsprit kept the bow pointed downwind. The rain quickly obscured everything beyond the dim outline of the boat. He couldn't see anything beyond the bow, but he remained standing in the companionway, partly sheltered, ready for anything.

By 0230 the wind had risen to gale force. The shrieking howl in the rigging was augmented by the surflike noise of waves that rose and broke astern, a tumbling froth sweeping the decks. But after the first edge of the front had passed, the rain had lightened and there were holes in the clouds through which Knox-Johnston could now see stars. He went on deck and stood on the self-steering gear's tubing, holding tight to the rigging as the boat rolled heavily from side to side, and peered into the dark. He knew, as sailors come to, not to stare hard but to let the faintest suggestion of form and light in at the periphery of his gaze. He was hoping to spot Centre Island light again. He didn't know if he had passed it or was bearing down it. The island was surrounded by rocks. The wind was driving rain and spray from astern over the boat, drumming into his back hard enough to make it feel like hail, so he could see nothing behind him. He looked forward and to port and starboard and hoped.

After half an hour he noticed a periodic whitening in the

clouds to the north. It matched the pattern of Centre Island light. It seemed close, so he relashed the tiller a little to port to point the bow far enough away, he hoped, to clear the rocks southeast of the lighthouse. He stayed hanging in the rigging for another hour, hands growing numb, watching for the breakers, knowing that if he saw them at all it would probably be too late. Finally he decided he must be past the rocks and went below to make himself Irish coffee and smoke a cigarette.

Soon it began to grow light outside, but daylight, when it came, offered little help in fixing his position. Visibility was about a mile, and all he could see was rough water. He feared he would sail down the strait past Bluff without seeing it, so he adjusted the tiller again, putting the boat back on a course to close the land.

Staring into the misty gray cloud-swept air he constantly saw phantom apparitions of land. But at 0730, one of these resolved into an unmovable shape, dead ahead. It was the sailor's most exquisitely fashioned nightmare: land, a lee shore in a storm, and *Suhaili* was being swept fast toward it. The wind at his back was now so strong that Knox-Johnston seriously wondered if the mast or sails could bear the strain, but he had no choice. He had to set the mainsail and try to claw away to windward. As he raised the mainsail, he could feel *Suhaili* shuddering and burying her decks under the strain. He set her on a course that he hoped would clear the looming land, and then he began to haul in the big blue warp and sea anchor, which were holding the boat back, keeping him pinned near the shore. These had tangled together into a vast, thick, 720-foot-long braid, impossible to untangle, dragging heavily astern as *Suhaili,* driven by the gale, tried to leap and pull ahead. Desperation and determination helped him pull the tangle aboard. By the time he got this heap of line into the cockpit, the cliff to leeward appeared much closer. He could even see the spume from waves breaking at its foot. But it came to an abrupt end in a headland farther down the coast, and Knox-Johnston wondered if this was, in fact, Bluff.

Suhaili drove through the big seas and water exploded over her with every wave. But with the warp and sea anchor aboard, she began to haul away from the land, and beating to windward, she held her course nicely. Knox-Johnston's back and arms were aching, his hands red and throbbing. He went below to make himself coffee. The motion inside the cabin precluded any possibility of boiling water, but he had taken the precaution of filling his thermos with hot water during the night. As he held a hot mug in his raw hands and smoked a cigarette, he was aware of a feeling of euphoria. He and his boat were being subjected to their most supreme test, and so far they were getting through it.

When he went back on deck a few minutes later he noticed that the seas were down, although the wind was still fully gale-force. He realized he must be in the lee of Stewart Island, and closing with land. Soon, indeed, land appeared close to starboard, and he tacked and headed north.

A ferry rolled heavily out of the mist ahead. It altered course and steamed over to *Suhaili,* coming close enough for Knox-Johnston to shout above the noise of the wind to the crew. They knew who he was and told him that Bluff lay 9 miles north.

At 1030, Bluff appeared ahead, identifiable by its lighthouse. But out of the protecting lee of Stewart Island, *Suhaili* was once again exposed to the full force of the storm, and now the tide had turned and was setting hard east through the strait. Although Knox-Johnston tried beating to windward again, *Suhaili* was swept away like flotsam. He started to reef the sails to slow down, but the mainsail halyard chose this moment to jam at the top of the mast and the best he could do was raise the boom and tie it and the flapping sail to the mast. He streamed his tangle of warps once more and headed off downwind. The storm continued all day, blowing at force 10 (48 to 55 knots) until early afternoon. By dusk he was clear of Foveaux Strait. There would be no going back now. Still hopeful of meeting Bruce Maxwell, he adjusted the tiller until *Suhaili* was now being blown northeast, paralleling the coast. He went below and slept until daylight the next morning. The storm was over.

That evening he sailed around Tairoa Heads into Otago Harbour, 130 miles northeast of Foveaux Strait. The "harbor" turned out to be an inlet sided with green hills and sand dunes, not the industrial port he had expected where he might have attracted attention and sent word to Maxwell. He blew his foghorn off the signal station but got no response. As he sailed slowly back out of the channel, the wind turned light and flukey around the headland cliff, and while he was trying to tack away from the rocks ashore, *Suhaili* stopped moving—she had run softly aground.

The bottom appeared to be sand, and the tide was ebbing. Knox-Johnston went below and brought an anchor back on deck. He tied a line to it, stripped off and jumped overboard. He walked along the bottom toward deeper water—away from the shore rocks—and continued walking when the bottom deepened and the water rose above his head, bouncing up for air every few seconds. Finally he sank the anchor into the soft bottom and swam back to the boat. It was his first view of *Suhaili* from a distance in 5 months: she looked dirty and rust-streaked. She began to heel over as the tide dropped.

A man called down to him from the cliffs above. He said he would send help, but Knox-Johnston was emphatic that he didn't want help—it would disqualify him—he would get himself off when the tide came back in. He asked about Bruce Maxwell and the man on the cliff said he'd try to find him.

A few minutes later a small motorboat and a crayfishing boat came by. The men aboard all knew who Knox-Johnston was and told him that Bruce Maxwell had been rushing up and down the coast looking for him. He lit a cigarette and sat on the cabin roof talking to them, enjoying the novelty of their company and the first stillness he had felt in 159 days. The Kiwi boats had radiotelephones and soon they heard that Bruce had been found and was on his way. While they talked, Knox-Johnston hauled himself up his mainmast and freed the jammed halyard. It grew dark. The tide turned and began to flood. At 2300 the keel began to bump on the bottom; he winched in the anchor line and *Suhaili* was soon floating free. There was no more water in the bilge than usual; she appeared unharmed by the grounding. The crayfishermen motored off to try to find Maxwell, and Knox-Johnston went below to make a quick supper. Soon there was a shout.

"What the hell are you doing here?" said Bruce Maxwell.

The first thing he had to tell Knox-Johnston was that one of the *Sunday Times*'s rules—no material assistance of any kind—had been taken to mean no mail. Where it had stretched its rules to allow the involvement of any sailor, known or unknown, in its race, and allowed them to proceed to sea without any inspection of their boats or equipment, the *Sunday Times* had finally found something to be stringent about. Knox-Johnston was outraged. Sailors have always hungered for news from home, and he would have happily disqualified himself on the spot if Maxwell had any mail to give him. The reporter, knowing this, had brought none.

Instead, he told him that his family was well, that Bill King was out of the race, that there were three new competitors: Carozzo, Tetley, and Crowhurst. None of them posed much threat, but Maxwell told him that Moitessier was fast closing the gap between them, and if they both maintained their present average speeds the race would end in a photo finish.

Maxwell left to find a phone and call London, intending to come right back, but the wind rose and Knox-Johnston, tenuously anchored, with rocks to leeward, decided not to wait for him. He raised his sails, tacked out over his anchor, hauling it aboard when he felt he was clear of the rocks. He shouted good-bye to the crayfishermen and headed for the open sea.

He wanted to get a move on. The Frenchman was coming.

S

19

WITH FRESH NEWS of Robin Knox-Johnston, the *Sunday Times'* navigation experts predicted a "neck-and-neck" finish between him and Bernard Moitessier.

The Frenchman, they calculated, had by then covered 14,000 miles at an average of 128.4 miles a day, against the Englishman's 17,400 miles at 98.3 miles per day. The experts gave the edge to Moitessier, who, they thought, could reach England on April 24, with Knox-Johnston arriving six days later, on April 30. With shiploads of uncertainties, anything could change these dates. The race was either man's to win or lose.

For the first time, the newspaper described the young Englishman in heroic terms: the "courageous . . . ruggedly handsome" sailor was proving himself to be in the same class as Chichester. Warm praise indeed.

The *Sunday Times* was positioning itself to recognize a possible victor, but it was still hedging its bets. Moitessier was "a cunning navigator" who had come daringly close to the Cape of Good Hope to save sea miles and was planning to sail far south of Australia for the same reason.

Italy's Chichester, Alex Carozzo, was already out of the race. He had spent almost a week at his mooring in Cowes preparing

Gancia Americano for sea before casting off, and had sailed as far as the Bay of Biscay, off the coast of France, when he began to vomit blood. He spoke with a doctor by radio, who diagnosed ulcer trouble, which Carozzo had suffered before. The doctor prescribed a bland diet and advised Carozzo to give up if the bleeding continued as there was a risk of hemorrhage. Carozzo sailed on, hoping for an improvement. His giant yacht was sailing fast and he was eager to keep going. But the bleeding didn't stop. Off the Portuguese coast he radioed the *Sunday Times* that he was making for Lisbon and emergency hospital treatment. A Portuguese air force search-and-rescue plane spotted him, and a pilot boat towed him into Oporto. From his hospital bed in Oporto, Carozzo told a reporter that he thought his ulcer had been aggravated by the strain of his intensive preparations for the race. He had read that Bill King was planning to try again the following year, and he was thinking of doing the same. He hoped the *Sunday Times* might be interested in a second race.

Although they were far behind Moitessier and Knox-Johnston, the *Sunday Times* did not discount the potential of the two trimaran skippers, Nigel Tetley and Donald Crowhurst, who "could surprise us all."

Crowhurst in particular. After averaging 60 miles per day during his first month at sea, a slower speed than any other competitor for the same stretch of ocean, the *Sunday Times* reported a sudden surge in his performance. During his fifth week he had covered 1,020 miles, at an average of 150 miles per day. This was the sort of dazzling speed he had promised with his revolutionary boat.

Nigel Tetley was not having as much luck. A week of light winds around the island of Trindade got him 473 miles—67 miles per day. He was still not able to transmit over his radio, though he could pick up stations as far away as Holland quite clearly. He worried about Eve and his family, who were doubtless concerned about him. Yet life aboard *Victress* was comfortable and civilized to a degree unimagined aboard his competitors' boats. He dined on cockles, prawns, asparagus, lobster, Polish sausage, smoked salmon, mushrooms, Pacific Coast oysters, roast pheasant, while listening to Respighi, Boccherini, Delius, Sibelius, Kodály, Dohnányi, and Saint-Saëns. After seven weeks at sea, he came to the end of his Music for Pleasure repertoire and wrote conscientiously in his logbook, "I can say here with all honesty that the majority have given me great pleasure and contributed in no small way to my peace of mind."

So far his voyage was slow and steady, unspectacular, and seamanlike. No great terrors of weather had been met. There were no apparent problems, other than the nagging frustration of not being able to raise anyone on the radio. It was almost dull.

Victress was performing well, though she was not proving as fast as he had hoped. Small problems were showing themselves: the molding strips covering the hull-to-deck joints had been breaking off, cabin windows were leaking. But these were ordinary signs of wear. Even during a long summer's sailing season, when most boats will actually spend more time at a dock or at anchor, such problems will develop. *Victress*'s constant exposure to wind and sea was having the effect of steady, but so far not alarming, attrition.

But a routine check below the floorboards of *Victress*'s outer hulls revealed them to be awash with seawater: 10 gallons in the port hull, 70 gallons in the starboard; he had been carrying 800 pounds of extra weight. This had not been apparent before because to check below the floorboards he had to remove all the gear piled on top of them, not a job he could do often, and then only in fine weather. This showed steady leakage that he would have to keep an eye on.

On November 19 he at last made patchy contact with Cape Town Radio, and over the next few days transmissions became clearer. He learned that Bill King had dropped out, and he heard that Robin Knox-Johnston was apparently in some trouble off New Zealand. These reports was unnerving to hear, but his weather, in approximately the same area where *Galway Blazer* had been rolled over and *Suhaili* had suffered her first knock-down, remained almost too benign, resulting in frustratingly slow speeds.

Early in December he passed south of Cape Town and sailed into the Indian Ocean. He still didn't quite venture into the Roaring Forties, but sailed east, keeping several degrees north of the fortieth parallel, the northerly limit of the Southern Ocean. Nevertheless, he got a taste of Roaring Forties weather. He experienced the strongest gale of his voyage so far on December 11, when he estimated the wind at force 9 (41 to 47 knots). He steered downwind, reducing sail until *Victress* carried only a storm jib sheeted flat.

After a long night sitting in the wheelhouse spinning the wheel from side to side to keep the stern before the waves, Tetley was cold and miserable and wondering what exactly he was doing there. He had every reason to believe Moitessier when the Horn-tempered Frenchman had advised him to run before the wind under storm conditions; but now, cold, hungry, fed up, and unable to leave the wheel if he did keep running, he decided to drop all sail and lie ahull. But this seemed to suit the trimaran; its shallow purchase on the water gave it little resistance to the wind and waves, and it rode easily, like an albatross, Tetley wrote in his log, beam-on and skidding away sideways before the waves. Below the motion was easier, and he was glad to make himself a breakfast of coffee and Irish stew.

Because of his latitudinal fence-sitting, staying just north of the Southern Ocean, Tetley was frustrated by a mix of light and heavy winds. On December 15, after a fast, bumpy night during which he carried full sail, he recorded his best run of the voyage so far, an impressive 202 miles for the last 24 hours. The next morning he was becalmed, and the wind, when it returned, remained light for several days. He carried as much sail as possible, hoping to make up for the slow days, but this was clearly not the latitude at which to maintain speed, and he knew it.

Tetley carried the same pilot charts and sailing directions from the British Admiralty publication *Ocean Passages for the World* that all the other sailors had aboard their yachts, which show the routes and pathways in the sea where the most advantageous winds, currents, and conditions would be found. Admittedly, these directions were not intended for small yachts; they were written for and based on the observations of large sailing ships—the tea clippers, the four-and five-masted grain ships—which had always sought out the strongest weather in order to make the fastest voyages. Yachtsmen following in their wakes using these big-ship guidelines faced the task of plotting a course somewhere between prudence and the prescribed fastest routes which lay to the far south in the high Roaring Forties and the Furious Fifties, where Antarctic fogs hide drifting icebergs. This was a highly individual choice. Knox-Johnston tried to keep to the fortieth parallel and found it appreciably windier than just a few hundred miles to the north, where Tetley was staying. Moitessier had plunged deep into the Southern Ocean. In addition to *Ocean Passages* and the usual charts, Moitessier had with him fifteen letters from old "Cape Horners," men in their eighties, dim of vision but still sharp of memory, to whom he had written asking about Southern Ocean conditions. From their accounts, and his own earlier voyage around the Horn, he believed 1968 was proving to be an exceptionally warm and benign year in the Southern Hemisphere, and he made good use of the winds in the far south.

Holding to his more northerly latitude, probably out of concern for *Victress*'s light construction, Tetley was plagued by light winds and poor progress. Now that he had regained radio communication, Cape Town Radio was able to patch him through to contacts at the *Sunday Times*. They were disappointed by the lack of drama in his voyage. Had he fallen overboard, perhaps, or had anything else exciting happened to him? they wanted to know.

The dullness and monotony of his voyage was bothering him too. With winds more often too light than too strong, he spent hours staring at the flat sea, frustrated by his poor progress. Windless calm at sea is the sailor's true bane, worse than any storm, which at least provides him with steady physical and mental activity. Becalmed, he can do nothing to bring wind but wait for it. He can read or write or listen to the radio, but he will also idle away hours staring at the pretty sea in every direction, trying to stave off the creeping, irrational fear that the wind may never return again. He is repeatedly forced to acknowledge its stretching absence as he records conditions every few hours in his logbook, writing, "becalmed" again and again, while the handwriting betrays the mounting tension as the pencil pushes deeper into the paper, the word scrawled off with anger. Time and geography grow perversely elastic. The quiet that comes with a true calm at sea is, like the reverse picture of a photographic negative, pronounced and conspicuous for its inversion of the normal. No wind or water noise, no breath of air across the ear. The boat nearly still, heaving slightly on the ghost of a swell, will provide the only real sounds: the rolling pencil on the chart table. The aural void then fills with the rhythmic rushing of one's own blood, reminding the listener of his mayfly speck of mortality. He goes below to flip on the radio again. To be becalmed during an ocean crossing that may take a total of three or four weeks from shore to shore is one thing; but in a race around the world, the enormity of which can boggle the mind at the happiest of times, lying utterly still on the perversely flat sea while being certain that one's fellow competitors are being blown along at speed elsewhere, is hard to bear. It can be unhingeing.

Not a chatty logbook writer like Knox-Johnston or a soul-barer like Moitessier, Tetley nevertheless had admitted in his log on November 27 to a growing depression. "The further I go, the madder this race seems. An almost overwhelming temptation to retire and head for Cape Town is growing inside me—the cold finger of reason points constantly in that direction."

But the very next day, he believed he had cured himself of his melancholy. He had been neglecting his daily milk-and-vitamins drink, he wrote, which, once taken up again, magically lifted his mood of depression. It's possible that Nigel Tetley's low spirits were revived by vitamins and minerals, as he believed, but there would come a time when no amount of milk would save him.

20

On December 15, the *Sunday Times* reported that Donald Crowhurst's speed had increased again, dramatically.

His name—largely ignored through the long months leading up to the start of the race—now led the news of the fleet.

Crowhurst World Speed Record?

Donald Crowhurst, last man out in the *Sunday Times* round-the-world lone-man yacht race, covered a breathtaking and possibly record-breaking 243 miles in his 41-ft. trimaran *Teignmouth Electron* last Sunday. The achievement is even more remarkable in the light of the very poor speeds in the first three weeks of his voyage; he took longer to reach the Cape Verdes than any other competitor.

Is the 243 miles in one day a world record? Captain Terence Shaw, of the Royal Western Yacht Club, Plymouth, says: "If anyone has bettered it, and I doubt if they have, they can come forward with a counter claim."

The 36-year-old Bridgwater man now feels he has a fair chance of being the first home. . . . His message

> ended: "I have been listening to the European money crisis on the BBC and you can tell the *Sunday Times* that if I win they can pay me in deutsche marks."

Crowhurst had cabled the news of his record-breaking run to Rodney Hallworth on December 10. The short cable also gave his daily mileages for the preceding four days: Friday (December 6) 172, Saturday 109, Sunday 243, Monday 174, Tuesday 145. He made a radiotelephone call to Hallworth the following day, Wednesday, during which he made the comment about the deutsche marks, which Hallworth could not resist passing on. It made great copy.

Captain Craig Rich, an instructor at the School of Navigation in London, who was advising the *Sunday Times* on navigational matters, had been keeping a chart of the competitors' progress. Their speeds had all been generally consistent from the very beginning of the race. After his very slow start, Crowhurst's performance was the conspicuous exception. It had been fluctuating wildly, and Captain Rich now told the reporters that he was surprised by this record-breaking run.

Sir Francis Chichester was skeptical. He telephoned the *Sunday Times* to say that Crowhurst had to be "a bit of a joker" and that his claims needed close examination. But there was no way any examination could be made until Crowhurst returned to England. Until then, they would simply have to take his word for it.

The *Sunday Times* and the rest of the national Sunday and daily papers reported the record-breaking run, and Crowhurst's cheeky comments, with gusto. The report cranked up the excitement of the race and increased speculation about its possible outcome. It was good news for everyone, and the sort of fanfare Donald Crowhurst had always hoped and expected for himself.

After concluding that he could not sail around the world in his faulty boat, Crowhurst had sailed on, unable to face going home.

He made radiotelephone calls to Clare and to Stanley Best, but to neither did he voice his dilemma. He spoke of the boat's problems matter-of-factly, and of his efforts to fix them, and gave every indication of an unswerving intention to keep going.

Yet between November 19 and 21 he slowed almost to a stop, sailing listlessly in a small circle north of Madeira. He got out his Admiralty pilot book of sailing directions for Portugal and its islands, and from its written descriptions drew a detailed map of Funchal harbor: he was considering making for port there and putting an end to his ordeal. Then he changed his mind. On November 22 he began sailing at speed again, steering southwest, avoiding Madeira by a wide berth.

Sometime around the beginning of December, he came to a fateful decision. The normal daily, sometimes hourly, comments confided by all navigators to their logbooks—a ship's primary record of events, which Crowhurst too had made from the very beginning of his voyage: the sea state, frustrations with the boat, anxieties, problems, and successes—had by now vanished from his logbook. By December, he no longer cared to muse to himself on his bloody awful options. His logbook entries were reduced to the mathematical workings of his celestial sights: page after page of neat, penciled calculations. No word of inner torment.

On December 6 he opened a clean logbook—although the first was only half full—and in this second logbook he began keeping track of his navigational progress, his actual positions, a continuation of the record started in the first book. In the first logbook, from December 6, he began plotting a second record: a carefully detailed series of fake positions, each day placing *Teignmouth Electron* farther and farther from his actual location.

The business of celestial navigation is intricate but not difficult. It is part science and part seat-of-the-pants instinct, and the latter part is what makes it interesting and breeds pride and vanity in navigators. It begins with a sextant measurement of the angle between the navigator, the horizon, and a celestial body—sun, stars, or moon. Skill in the use of the sextant is acquired gradually with time and practice. Accuracy with the instrument depends on one's familiarity with it under a wide range of conditions and the experience to judge the quality of each reading. A navigator's proficiency with a sextant is like that of a gunslinger's with a pistol, something that over the course of time and many different situations breeds an instinctive ease of handling and nicety of result. It is said that a navigator's second thousand sights show a considerable improvement over the first thousand.

The math is the straightforward part. The sextant reading is taken through a series of corrections for height of eye above the sea surface, atmospheric refraction, time of day at one's own location, and Greenwich mean time; that altered set of numbers takes the navigator into a book of gloriously precomputed tables of spherical trigonometry that are no harder to use than a telephone book.

That is what Donald Crowhurst—and all the Golden Globe sailors—did to determine true positions, and which he continued to do to know where actually he was. But to calculate the second, fraudulent series of positions, based on imaginary extrapolations of geography and mathematics, is a formidably daunting exercise that would stump most honest navigators. Crowhurst, an able mathematician with a ready grasp of both the technical and the abstruse, was up to it. But it made a lot of work for him and added immeasurably to the deepening stress of his situation.

He began preparing the fake record for eventual scrutiny. It had to appear seamless, so from December 6, in the logbook he had been keeping since the beginning of the voyage, he wrote his false positions, and the calculations for them, and specious, salty descriptions of his day.

Crowhurst's working methods, bred and polished in the scientific laboratories of the military and private electronics companies, were neat and meticulous. He made notes for himself about everything; even before he had begun his deception, he wrote outlines of what he wanted to say when making innocuous radiotelephone calls to Clare. By the time he cabled Rodney Hallworth on December 10 with news of his record run, he had a neatly transcribed table of fake and actual positions for each day worked out. The remarks alongside the fake positions were descriptions of shipboard routine, difficulties, and food he had prepared, much of it written in the sort of gruff heroic tone used by Chichester in his book Gypsy Moth IV *Circles the World.* Crowhurst had read that book over and over, and now he was creating his own mythology.

On December 12, *Teignmouth Electron* was becalmed 400 miles north of the equator. During the morning, an ordinary tropical rain squall swept over the yacht, and its winds of perhaps 20 knots damaged a part of the Hasler wind vane. This became fodder for a heroic report for the media. The next day, December 13, Crowhurst cabled Rodney Hallworth that a 45-knot line squall had smashed the wind vane, but he thought repairs were possible. Four days later, while he was still 180 miles north of the equator, he cabled Hallworth that he was "over" the equator and sailing fast. On December 20, he cabled again saying that he was off the coast of Brazil and averaging 170 miles daily.

His true mileage for that day was 13.

He was stuck in the Doldrums, a band of calms or light and flukey winds, generally 600 to 800 miles wide, which lies roughly along the equator between the trade wind belts of the Northern and Southern Hemispheres. Wind-ship sailors have always hated this area, where they can drift for days while making no progress. Yachts—unless they are racing, when use of an engine is cheating—have always tried to carry enough fuel to power through the Doldrums. Crowhurst's phony reports gave no evidence of his being slowed down at all here, as were the other competitors. He appeared, to those at home, to have experienced no Doldrums at all, but amid the press excitement generated by his progress, no one, except Craig Rich and Sir Francis Chichester, wondered at his marvelous luck.

How much of his deception was, at this stage, real intent, or still simply a feasibility study, the sort of thing Crowhurst's engineer's mind would readily tackle, is uncertain. At least one other Golden Globe competitor admitted to toying with the same idea.

> I even considered the idea of simply resting in the sun for a year and then returning home to say that I had been all the way round the world. . . .

This popped into John Ridgway's mind as he grew unbearably lonely and anxious and thought about giving up, and the shame of abandoning his project, of disappointing all the people who had helped him. Out of sight, in a remote corner of any ocean or up a jungle river, such a thing might have been possible. But Ridgway quickly discounted it.

> First, I doubted if it could be carried off, too many people would see through the story. Second, and more important, it would not be possible for me to live with such a fabrication.

But Crowhurst's cabled positions were not simply wishful chunks of mileage that sounded good. He had started working out consistent dates and positions for his fake voyage that stretched out for weeks ahead. He plotted them directly onto a routing chart. What began as a discrepancy of a few hundred miles between his actual and fake positions quickly grew to projections of thousands of miles away. By drawing a line between these points, connecting the dots on the chart, he could then tell at any time on any day where he was supposed to be and what his course was. This would be a necessity for keeping Hallworth and the world informed of his progress.

Yet at the same time he was also making a detailed map of the harbor at Rio de Janeiro. Like the one he had drawn of Funchal in Madeira, this was full of information taken from small-scale charts and his Admiralty pilot books, showing lights, landmarks, navigational hazards. This is what a navigator must do when making for a port he hasn't planned to visit, and for which, therefore, he is carrying no large-scale charts aboard his vessel. There would be no other reason to make such a harbor chart.

Crowhurst would certainly not have thought of putting in at Rio on the quiet to make repairs and then returning to the race. His boat would instantly have been seen and visited by harbor police and immigration officials. The only reason for going into Rio would have been to give up.

Whatever torments of indecision still plagued him, he soon sent another cable to Rodney Hallworth. This did not give a sailorly latitude and longitude position; few of Crowhurst's cables did. Instead, they invariably named or suggested a location—"off Brazil"—that implied a position from which tremendous progress could be inferred. On this cable he wrote that he was sailing "towards Trinidade." He meant Trindade, which both Moitessier and Tetley had sighted. This was even getting ahead of his fakery, for it was 350 miles south of the false position he had worked out for December 24. Somewhere, either in the radio transmission of the cable to England (all the cables were sent by Morse code, at which Crowhurst was highly proficient) or in its transcription, by the time the cable reached Hallworth at his Devon News office, the *e* had fallen off Trinidade. Hallworth knew Crowhurst didn't mean Trinidad in the West Indies, so, with a hazy but convenient grasp of geography and unbounded optimism and pride in his client, he remembered another island somewhere down there in the South Atlantic, where Donald surely was now that he had crossed the equator, which had figured in the reports of other Golden Globe racers: Tristan da Cunha. That's where Donald was, Hallworth realized. Tristan da Cunha. About 2,500 miles beyond Crowhurst's carefully calculated fake position and 3,000 miles from his actual position—and, better still, at 38 degrees south, at the very edge of the Roaring Forties.

Hallworth's boy was doing sensationally, and he made sure the press knew it.

S

21

At 7.51 a.m. EST on Saturday, December 21, 1968, *Apollo 8* took off from Cape Kennedy, heading for a drive-by on the dark side of the moon. It was the *Apollo* program's final run-up to the epochal voyage that President John Kennedy had decreed must be made before the decade was out: the landing of a man on the moon and returning him safely to Earth.

Apollo 8 carried three brave astronauts: Frank Borman, 40, James Lovell, 40, and William Anders, 35. Their journey of half a million miles, there and back, and ten orbits of the moon, would be made, if all went well, in 6 days. They sped toward the moon at 24,000 miles—a distance equal to a little more than the circumference of the earth—per hour.

This was naturally the front-page story in the *Sunday Times* on December 22.

Deep inside the paper a single column gave the latest details of the much dodgier business of getting a man alone in a sailboat around the world in 10 months.

Robin Knox-Johnston, who had not been heard from since tacking out of Otago Harbour on November 20, was thought to be halfway across the Pacific, in a region where gale-force winds occurred with "disconcerting frequency." His sponsor, the *Sunday Mirror,* expected him to round Cape Horn sometime in early January.

Of Moitessier there was fresh news. He had been sighted off the coast of Tasmania four days before, on Wednesday, December 18. The Indian Ocean had slowed him down: he had covered the 6,000 miles between the Cape of Good Hope and Tasmania at an average of 100 miles per day. Nevertheless, his overall daily average was 128.4 miles, and the gap between him and Knox-Johnston was closing by 210 miles per week. It still looked to be a photo finish in England some time in April.

Nigel Tetley had made contact with Perth Radio, Australia, even more recently, on Friday, December 20, giving a position in the middle of the Indian Ocean, not far from Amsterdam Island and St. Paul's Rocks. He had covered a fast 185 miles in the previous 24 hours. His total distance sailed was now 9,900 miles at an average of 100 miles per day. Tetley's steady, dogged, seamanlike progress provided little in the way of exciting copy, so the *Times* noted that both Amsterdam Island and St. Paul's Rocks were uninhabited but stocked with depots containing clothing and provisions for castaways.

Rodney Hallworth had not yet turned Trindade into Tristan da Cunha, and the *Sunday Times,* while reporting that Donald Crowhurst had crossed the equator, gave only his last known position, several hundred miles north of the equator. Despite his surging progress, Crowhurst's daily average was still far below Tetley's, at 79 miles per day. But his every communication was more exciting, and the newspaper was able to relate the damage done by the 45-knot wind that had smashed his self-steering gear and a jib pole.

It was a week before Christmas when Varley Wisby and his two sons, fishing off the southwest coast of Tasmania, saw a red-hulled ketch coming straight for them. On deck a lone man was flashing a small mirror, catching the sunlight—he was signaling to them. They steered for the sailboat, then slowed, maneuvering their fishing boat carefully until it ranged alongside the ketch, close but not too close, matching its speed.

The sailor, a Frenchman, who looked emaciated beneath a filthy wool sweater and baggy black pants, his hair and gray-streaked beard as long and wild as a yogi's, tossed a metal film can across the water to the Wisbys. He told them he was in a yacht race and asked them to give the can to someone who would pass it on to the *Sunday Times* in England. Varley promised to give it to the commodore of the Royal Tasmanian Yacht Club when they returned to shore in three days' time.

The Frenchman asked the whereabouts of three other racers—Bill King, Nigel Tetley, Loïck Fougeron—but the Wisbys knew nothing of them. One of Varley's boys had heard something about an English yachtsman who had sailed past New Zealand without stopping. When? the Frenchman asked eagerly. The Wisby boy didn't know. He'd heard it on the radio sometime in the past month maybe. The sailor thanked them, turned his boat by fiddling with the little vane at its stern, and veered off toward open water. Varley and his boys watched him go.

Moitessier had sworn to himself he would not risk his boat and voyage again to send word to his family and the press, but he also hungered for news of his friends and rivals, and he had spent a long day and a sleepless night sailing through rain squalls in the Entrecasteaux Channel, off Hobart, Tasmania, in hopes of finding a boat. During the black night, the phosphorescence in the water was so bright and glowing that he repeatedly believed it to be breakers on a reef. He hove to half a dozen times to listen for any sound of danger. He was breaking all his own well-learned rules to be here, and all through the night he feared the price was waiting for him somewhere in the darkness.

But then dawn came and he found the fishing boat, delivered his can of film and letters and messages, and made his getaway without mishap. The sky cleared, the wind fell away to a breeze from the west, and he headed offshore, gliding close enough to a lighthouse to hear the ratcheting of a cricket ashore. Anxiety gave way to joy as he sailed into the Tasman Sea.

Moitessier had found a very different Indian Ocean to the one that had pummeled Robin Knox-Johnston and *Suhaili*. Calms and winds too light had been his portion. He spent weeks sailing slowly if steadily, spending a lot of time on deck watching the ever-present albatrosses, mollymawks, Cape pigeons, and shearwaters. He practiced yoga daily. He sat meditating for hours on *Joshua*'s deck, his long skinny legs easily pretzeled into the full lotus position.

Once into the Tasman Sea, however, the Southern Ocean winds found him again and his speed picked up. With sails reefed and water tearing past *Joshua*'s hull, he made daily runs of 164, 147, 153 miles. He now religiously listened to the BBC World Service for news that he had been sighted off Tasmania, hoping that this might prompt a mention of Fougeron, Tetley, and King, but he heard nothing.

Nowhere in Moitessier's writing is he as respectful and affectionate as with his sailor friends, with whom he formed his closest bonds. They alone, he believed, shared and understood what he felt and knew about the sea—they understood him in ways his wife, girlfriends, and children did not. He was well aware that the greatest sailorly skills could mean nothing on the wrong day at sea, and he was constantly anxious about the welfare of the three men with whom he had shared plans, techniques, and hopes at Plymouth. His daily hope of hearing any word about them went unsatisfied. Whether because of this, or his brief contact with the Wisbys, or because he was less self-sufficient than he liked to think, with the approach of Christmas Moitessier was uncharacteristically overcome with loneliness.

On Christmas Day he sighted the Cameron Mountains on New Zealand's South Island. In unusually clear conditions, they stood up above the horizon, 50 miles away. Usually spartan and routine with his food, Moitessier took pains to make himself a Christmas dinner. Into a pot went a smoked York ham, a can of hearts of lettuce, garlic, onions, a can of tomato sauce, a quarter of a can of camembert cheese.

Still he felt blue. He missed his friends and his family. And to torture himself, he remembered with remorseful detail a rat he had killed years before in Tahiti. He had found it on board and caught it by jamming it against the floor with a book. As he put a stone to his slingshot and took aim, the rat had given him a look that was haunting him still. Moitessier knew how the rat had felt: when the Japanese captured Saigon near the end of World War II, he had been imprisoned with his family. One day a Japanese guard came into the 20-year-old Moitessier's cell intending to kill him. He raised his pistol, but they locked eyes until, inexplicably, the guard lowered the pistol and walked away. Now, years later, tenderized by solitude, Moitessier wished he had spared the rat.

He drank away his guilt and grief and loneliness with a bottle of champagne given to him by *Joshua*'s designer, Jean Knocker, and went to bed with *Joshua* ghosting again over a calm sea.

Two days later, the animal kingdom reappeared with an unequivocal message that the business with the rat was forgotten. The west wind was freshening and *Joshua* was sailing fast, passing south of Stewart Island (which Knox-Johnston had passed to the north during his storm in the Foveaux Strait). Moitessier's dinner was growing cold in the pressure cooker on the stove because he wanted to pass the longitude of South Trap, a dangerous outlying reef below Stewart Island, before relaxing his vigilance and eating and sleeping. South Trap would mark his entrance to the Pacific Ocean, the last rocky obstacle between him and Cape Horn. He hopped up and down between deck and cabin, listening to and watching the sea, tweaking sheets for speed, rolling cigarettes below. *Joshua* sped on, steered as always by her vane gear.

In the afternoon dark clouds obscured the horizon to the north, where he might have seen Stewart Island, and a large school of porpoises, perhaps a hundred of them, appeared around the boat, whistling and clicking, turning the water white with their breaching and splashing. Usually, these "playful" (we anthropomorphically like to suppose) creatures swim alongside a yacht, criss-crossing singly or in synchronized groups in front of the bow wave. But that afternoon they gave Moitessier a show he had never seen before.

A tight line of twenty-five porpoises swam abreast off his starboard side, rushing from stern to bow, and then veering off sharply, always to the right. Again and again and again, more than ten times, they regrouped and made this same maneuver, while the rest of the school behaved in a manner Moitessier construed as nervous: they moved erratically, they beat their tails on the surface, they created pandemonium around the fast-sailing *Joshua*. All the while, a single platoon continued its streaking, abrupt right-hand-turn maneuver. *To the right. To the right.* Moitessier watched astounded.

Finally, instinctively, he looked at the compass, something he had not done in a while with the wind vane doing the steering. The west wind had shifted into the south without his noticing it, and *Joshua* was racing north, not east, toward the reefs of South Trap. Normally a shift in wind will alter the wave patterns of the surface, and very soon have them running at an angle to the older swell, a visible alteration of the sea state, and immediately felt by a sailor aboard a boat. But that afternoon there was, unusually, little or no swell, and Moitessier, not for the first time in his wreck-strewn life, had been fooled. He altered course to starboard, to the east—to the right, the direction of the porpoises' abrupt turn.

Their behavior changed immediately. Their nervousness, their disruption of the sea surface, disappeared. Now they swam in their usual playful way. And as Moitessier watched them, wondering but not wondering at all about what had happened, one large black-and-white porpoise leapt clear of the water and somersaulted twice in the air before flopping back onto the surface. Twice more it leapt out of the sea to perform its ecstatic double somersault. The school remained swimming alongside *Joshua* for another three hours, for a total of five hours, an extraordinary length of time for such a visit. At dusk, when he was well past South Trap, the porpoises disappeared.

On Monday, December 23, a strong gale overtook Nigel Tetley and *Victress*. At noon the thin PVC-coated wire connecting the steering wheel to the rudder parted. The same thing had happened 8,000 miles earlier, on October 11, when he had been off the Cape Verde Islands. This time he replaced it with heavier rigging wire, hoping it would last longer. *Victress* was doing well in the strong winds, being steered by her wind vane, but Tetley, still getting used to Southern Ocean conditions, went to bed fully dressed in rain gear and sea boots, ready to jump on deck for anything.

At nightfall on Christmas Eve the wind moderated, and Tetley began preparing his Christmas Day dinner. He had decided on a mushroom sauce for his roast pheasant and soaked some dried mushrooms. He tidied the cabin, baked bread, and got out his two presents to be opened in the morning.

The weather cooperated with his Christmas plans. The moderate wind settled into the west, and Tetley raised his twin running sails, which helped the wind vane, and the boat steered itself all day.

He opened his presents: a pewter tankard from Eve, a stainless steel comb from his son Mark. Strong metal from both. He drank sherry before lunch. He listened to a tape of Christmas carols from Guildford Cathedral.

He took a photograph of himself tucking in to his Christmas dinner. It shows that he has done his best to make the occasion and his surroundings as civilized as possible. The cabin table is decorated with his last two or three oranges, the contents of his last packet of nuts, some raisins and candy. There is his roast pheasant in his mushroom sauce on a small plate, Eve's tankard partially filled with champagne from the bottle that also sits on the table. There is little in the spacious cabin to indicate that he is far out at sea, or even (if one ignores the Indian Ocean chart partially visible in the foreground) on a boat at all. He looks for all the world like a man sitting down to his solitary Christmas dinner in a rather cramped but neat bed-sitter flat in London's Earl's Court, filled with dutiful Christmas spirit, and he couldn't look any lonelier.

Farther astern of him than anybody knew, Donald Crowhurst was struggling to hold onto himself over Christmas.

On Christmas Eve he recorded a rambling monologue on the tape recorder given to him by the BBC. He talked, trying for a Chichesterian tone, about the incessant work to be done aboard a yacht out at sea, but loneliness was on his mind and he couldn't keep it out of his Christmas musings. "There is something rather melancholy and desolate about this part of the Atlantic Ocean. . . . Not that I'm depressed or feeling sorry for myself by any means, but . . . Christmas . . . does tend to make one a little bit melancholy. And one thinks of one's friends and family, and one knows that they're thinking of one, and the sense of separation is somehow increased by the—by the loneliness of this spot. . . ." Then he played *Silent Night* on his harmonica, remarking afterward that it was a "melancholy" carol. He was a fair harmonica player, able to invest tunes like *Summertime* with a sad, bluesy feeling. He then tried to cheer himself up by playing *God Rest Ye Merry Gentlemen.*

Later he made a radiotelephone call to Clare. She asked him for a latitude-longitude position, which Rodney Hallworth badly wanted instead of the vague but suggestive geographics Crowhurst had been cabling him. He answered that he hadn't had a chance to take recent sights. Then—forgetting or ignoring his own well-calculated fake progress—he told her that he was "somewhere off Cape Town." He was well south of the equator now, but this wild exaggeration, thousands of miles beyond his actual or falsely plotted positions, was so impossible that it suggests a hopeless abandonment of any effort at pretense or reality.

Then he asked Clare how she was coping at home. The children, who had been following his progress on a chart taped to a wall, missed their father badly. There had been problems with the slow but supposedly ongoing manufacture and sales of his Navicator, the money from which they had both hoped would help keep her and the family going while he was away. In truth, hardly any money was coming in and Clare's position was approaching desperate; she would soon be forced to go on the dole. And two days earlier there had been a fire in the stable behind the house that had been his workshop. At this point, Crowhurst would almost certainly have sped happy and relieved straight for the nearest port if Clare had told him things were difficult or pleaded in any way for him to stop and come home. But she did not. She bravely mentioned nothing of these troubles to her husband, and he lied back to her that all was well with him. The impossibility for each of them of being honest with the other made for a strained call.

After they disconnected, Crowhurst could not leave the radio alone. He was desperate for human contact. He stayed up all night tuning into various shortwave frequencies, picking up news bulletins from around the world. He had been hoping for messages from Rodney Hallworth, Stanley Best, or even the town councilmen of Bridgwater, to all of whom he had sent Christmas cables, but he received nothing. At 0527 in the morning he recorded in his radio log: "Sighs heard."

From his night of loneliness and scrambled radio voices carrying snippets of dire news from around the world, he wove a tortured Christmas poem.

> Keeping a sort of watch on sails by night,
> Alone,
> The rigging sighs a sigh of cosmic sorrow
> For weeping doves that die maybe tomorrow
> On 12.7 x 10^5 irradiated olive trees.
> A sigh to fill a man's soul with melancholy.
> Waves! Sweep away my melancholy!
> My footstool's a 10 lb case of rice
> To the North-east 2.5 x 10^3 miles,
> 250 x 10^3 babies will slowly die, too weak to fuss

(Carbohydrate deficiency, they tell us
on 15.402 mHz)
Herrod, would you not solve overpopulation thus?
Please, be informed, there is a Santa Claus!

After his call to Clare, he steered southwest toward land, closing with the northeast Brazilian coast, coming within 20 miles of João Pessoa. Going on deck every now and then for a lookout, it's possible that he saw the loom of shore lights that night. He had not been this close to land since leaving England.

Then he altered course again, away from shore and people, heading southeast down into the South Atlantic.

Only Robin Knox-Johnston found within himself the makings of a merry Christmas. He began by feeling slightly put out at the idea of spending Christmas alone, but after he opened the whiskey, thoughts of his family and Christmases past soon had him laughing out loud. After two glasses he went out on deck, climbed on the cabin top, and belted out his own carol service. He ended Christmas Eve feeling "quite merry."

On Christmas Day he took care over the preparation of his dinner: he fried tins of stewed steak, potatoes, and peas, "cooked separately for a change," and made a currant duff (a lead-heavy English pudding traditionally made for schoolboys and sailors, embodying the worst of institutional English cuisine). At 3 in the afternoon—the time in England when the queen makes her Christmas speech, one of the special charms of Christmas for Knox-Johnston and his family—he drank a Loyal Toast.

That evening he tried contacting radio stations in New Zealand and Chile but was unable to get through. However, conditions were perfect for signals from AM stations in Texas, Illinois, and California to bounce off the atmosphere and reach him in the far South Pacific, and from one of these that night he first heard about the *Apollo 8* moon journey.

> It gave me food for thought. There they were, three men risking their lives to advance our knowledge, to expand the frontiers that have so far held us to this planet. The contrasts between their magnificent effort and my own trip were appalling. I was doing absolutely nothing to advance scientific knowledge. . . . True, once Chichester had shown that this trip was possible, I could not accept that anyone but a Briton should be the first to do it, and I wanted to be that Briton. But nevertheless to my mind there was still an element of selfishness in it. My mother, when asked for her opinion of the voyage before I sailed, had replied that she considered it "totally irresponsible" and on this Christmas Day I began to think she was right. I was sailing round the world simply because I bloody well wanted to—and, I realized, I was thoroughly enjoying myself.

His voyage might have been a small one compared to the *Apollo* flight, but he was driven by the same genetic impulse that powered NASA, and in his own way he was exploring the same boundaries of human endeavor. Furthermore, he was enjoying himself. That was Knox-Johnston's special adaptation and qualification. He was at home at sea.

John Ridgway was not. Nor was Chay Blyth. Both soldiers, brutally tough men, saw a circumnavigation as an ordeal to be endured, and both hated being at sea. Bill King enjoyed his voyage until he saw that others would probably beat him home; he lost heart long before he was capsized. Fougeron lacked the genetic impulse—he didn't want it badly enough. Crowhurst had devised a personal hell. Tetley was plugging away despite boredom and loneliness with a peculiar, dogged determination.

Only Moitessier equaled Knox-Johnston for the sheer pleasure he derived from his epic voyage. Only these two were really happy aboard their boats at sea.

S

22

IN 1842, AN AMERICAN NAVAL OFFICER, Matthew Fontaine Maury, took charge of the U.S. Navy's Depot of Charts and Instruments. There he began collecting and collating weather observations recorded by sea captains in their logbooks. He looked beyond the information gathered by naval ships to search out the logbooks and diaries of merchant ship masters. He found a treasure trove.

By the early nineteenth century, whaling ships—half the world's fleet being American vessels from two towns, Nantucket and New Bedford—were poking their bluff-bowed noses into every corner of the known and unknown world. As whales grew scarcer in the historic grounds long plundered by whalers, these heavy, unhandy ships ventured perilously deep into the Arctic and Antarctic, becoming in every respect vanguard explorers of the farthest reach of the earth's oceans. No less bold than explorers like Cook, they found little or no fame for their voyages, only the product sought by their industry. Assiduously, the whaling masters and their equally intrepid brethren aboard sealing vessels made notes and surveys, maps, drawings, and watercolor paintings of their newfound territories; and *every day,* throughout voyages lasting three and four years at a time away from their home ports, they wrote up morning, noon, and evening observations of the weather they were experiencing. "Heavy snow all morning with gales from W. snow and wind fell off in the pm. seas down with short fetch now in lee of Pt. Barrow. Fog in the evening, a warm breeze from direction of land to the S. So end these 24 hours."

Maury and his team gathered thousands of such observations and made of them wind and current charts covering the known world, with accompanying explanations and sailing directions. For the first time mariners had available to them reams of written and diagrammatic information about the seas they intended sailing to, rather than the useful but limited oral lore handed down from old salt to young.

Maury's work was the basis for the pilot charts and sailing directions now carried aboard all ships at sea. Today, instant weather faxes tell sailors what's coming their way, but at the time of the Golden Globe race, pilot charts were the main predictor of the weather and sea conditions a ship, or a yacht, would encounter.

Pilot charts are the visual opposite of land maps: the land masses at their perimeters are blank, but the sea spaces are crammed with information. Oceans are divided into a grid of near-squares, 5 degrees of latitude by 5 degrees of longitude (at the equator, this square is 300 by 300 miles; the longitude distances progressively diminish going farther north and south), and in each square is a "wind rose" giving the average wind strength and direction that will be found there, and the percentage of time when gales and calms can be experienced. They also show where icebergs may be encountered, the path of tropical and extratropical cyclones, atmospheric pressure, the direction of ocean currents, air and sea temperatures, magnetic variation, and the routes across the ocean taken by full- and low-powered ships. There are charts containing all this information for all the world's seas and oceans for each month of the year. They are the essential tool used by navigators, shipping companies, and lone yachtsmen to determine the optimum route across an ocean.

Thus a pilot chart can tell a sailor in the middle of the Pacific stretch of the Southern Ocean in January, heading for Cape Horn, what to expect in every 5-degree square along the way.

They will also guarantee his disappointment and frustration. All the information is statistically *averaged* from millions of observations, and on any given day the weather at sea, as ashore, can confound all predictions.

All through December 1968, Robin Knox-Johnston experienced winds in the Southern Ocean contrary to everything he had been expecting, and despite a break for Christmas cheer, this proved the most frustrating and anxious period of his whole voyage. In an area where the pilot charts, and every written account of passages made there, practically guaranteed the famous westerlies of the Roaring Forties, he was met with days and weeks of easterly winds blowing in his face, slowing him down, at times stopping him. Beating his way east against infuriating head winds, he was convinced that not far behind, and getting closer every day, "the Frenchman" was surfing downwind toward him, getting the favorable conditions touted by the pilot charts. It drove him mad. It made his British blood boil. It brought out his best English schoolboy xenophobia:

> December 9th, 1968 . . . Of all the lousy things to happen; easterlies in an area which is renowned for westerlies. . . . If the Frogs are meant to win—OK, but there is no need to torture me as well as allowing me to lose, and the Chinese could hardly have thought up a slower, more destructive method of torturing a person than this. . . .
>
> December 10th, 1968. No change still. I cannot make it out at all. . . . Perhaps if I decided to turn round and head back to New Zealand I'd get westerlies! . . .
>
> December 29th, 1968 . . . I just give up! Someone is going to have to rewrite the books! . . .
>
> December 30th, 1968 . . . Tacking north and south and making no progress at all, whilst somewhere to the west and probably not far away now, I'll bet the Frenchman is having beautiful westerlies.

For Knox-Johnston, who could not accept that anyone but an Englishman should be the first to make a solo nonstop circumnavigation, the specter of "the Frenchman" having all the luck with the weather and sailing hard up his backside was probably a great boon. From New Zealand onward, after Bruce Maxwell had told him of Moitessier's pace and position, Knox-Johnston's logbook makes frequent reference to this hard-pressing threat to his lead. It's unlikely that he would have driven himself and *Suhaili* quite so hard without it.

Tacking northeast and southeast, according to whichever slant was more favorable, adverse winds drove him north above the fortieth parallel. When he slept, he began to be tortured by a recurrent nightmare that his entire voyage was simply a qualifying preliminary to the real race, which would begin once they had all returned to England. Finally, when he reached 37 degrees south, he altered course in disgust and headed south, even a little west of south, for three days until he found the consistent westerlies he'd been looking for. He concluded that he had strayed too far north and should have turned south much earlier, and had wasted 10 days. For the rest of December he tried to keep to 48 degrees south. This was close to the dotted line on his chart that showed the northerly limit where icebergs might be encountered, but he felt he had lost so much ground to Moitessier that he would chance it.

Although it was now midsummer in the Southern Hemisphere, the weather remained grim. Gales and hail-hurling squalls regularly overtook *Suhaili*. The seas were still high, throwing the boat around, and Knox-Johnston inside it. Water still poured in through the closed hatches and around the edges of the cabin, and he lived and slept surrounded and clothed by constant wet and damp. He thought longingly of the tropics, which were, as they seemed, half a world away. He had by now read all the novels aboard, so when he huddled in his sodden sleeping bag reading for escape, it was now Bertrand Russell's *A History of Western Philosophy* that he waded through.

Early in January he added more south to his course, crossing into the fifties, and began his descent toward the Horn, which lay 1,500 miles away at 56 degrees south. As he neared South America and dropped farther and farther south, the weather brought authentic Cape Horn conditions: depression followed depression racing east, overtaking the boat, bringing gale-force winds that boxed the compass, epic crashing waves, and a steep drop in temperature. He had a heater aboard but stopped using it after January 13 because it was consuming too much kerosene. Wet clothes hung everywhere in the streaming cabin, and Knox-Johnston smoked, drank coffee, and upped his whiskey intake in a vain effort to stay warm.

In the early hours of January 10, he was awakened by the boat's violent motion as it lost way and luffed up into the wind. On deck he found the mainsail had split in half along one seam. It would take hours of sewing to repair, but he had an older mainsail aboard which he raised in its place—the older one now didn't look so bad compared to the frayed and damaged "new" one. Later, as he was making breakfast, the boat lurched and steaming hot porridge covered his hand. Large blisters rose over the burn, which he soon broke during hectic sail-handling, leaving the raw skin exposed to icy sea spray.

Repeated soakings and constant wet had taken their effect on his Marconi radio. It would still receive, but seemed no longer able to transmit. Daily he tried placing calls to stations he could hear, but none could hear him. The British newspapers reported that although he was expected off the Horn around January 12, he had not been seen or heard from since leaving Otago on November 21. Knox-Johnston heard on the shortwave *Voice of America* that the Chilean Navy was looking out for a "damaged" ketch battling toward Cape Horn.

On Sunday, January 12, the *Sunday Mirror,* Knox-Johnston's sponsor, took the optimistic line that he had rounded the Horn and was already heading north in the Atlantic. Falkland Islands Radio was listening for him. The *Sunday Times* speculated that he could still be in the Pacific, with "the hour of his trial" still awaiting him. In fact, that Sunday he was still 480 miles from the Horn, painfully breaking the repaired blisters on his burned hand.

On Monday, he noticed that the jib stay was stranding—its wires breaking and unraveling. In 40-knot winds and above, Knox-Johnston crawled out onto the bowsprit with wrenches in both hands—"hanging on with my eyelashes" as the boat lifted and plunged him in and out of drenching waves—while he unscrewed the bottom of the stay, lashed it back on board out of the way, and set the jib "flying," supported only by its own luffwire.

On Tuesday, the gooseneck of the main boom broke again. The metal casting where it was bolted to the fitting on the mainmast had sheared. The flying jib had flogged and opened a seam in its foot, and Knox-Johnston crawled to the bow to sew a few stitches in place.

Wednesday he made a crude but ingenious repair to the broken gooseneck, pirating a metal plate from the defunct self-steering system, sawing a slot into the end of the boom into which he bolted one end of the plate, bolting the other end onto the mainmast. He reinforced the jury-rigged end of the boom with a wrapping of fiberglass, and when that had hardened, rove around the repair two "turk's heads," decorative-looking rope whippings that were in fact stronger than the fiberglass. He finished the ropework as the wind and sea rose and waves began breaking over the boat. He was now 200 miles from Cape Horn.

Thursday, another depression brought storm-force winds of 50 knots. While he was sail-handling on deck, hail drove into the raw, weeping flesh exposed beneath the blisters on his burned hand.

Early on Friday morning, January 17, he spotted the lonely Horn outrider islands of the Diego Ramirez group 15 miles to the south. The drowned chain of the southernmost Andes appeared to the north a few hours later. The day was squally, but with clearing periods between the squalls, when Knox-Johnston could gaze at the rugged cordilleras capped with glacial ice and think about the immensity of his voyage and the home stretch awaiting him after rounding this last and greatest marker on the race course. The cold wind backed from the west-northwest to the west during the day, and dropped in strength through the afternoon until it blew at a gentle 5 knots by early evening.

At 1900 a brief rain squall passed over the boat and when it cleared, Cape Horn—a false cape, for it is really a small island with a sphinxlike profile, far less impressive in appearance than Gibraltar—was clearly visible to the north. He had already passed it.

"Yippee!!!" Knox-Johnston wrote in his logbook. Then he had a drink and opened his Aunt Aileen's fruit cake. It had spent 7 months wrapped in foil inside a cake tin and was in perfect condition. Also in the tin was a page of the *Times*. He had something new to read.

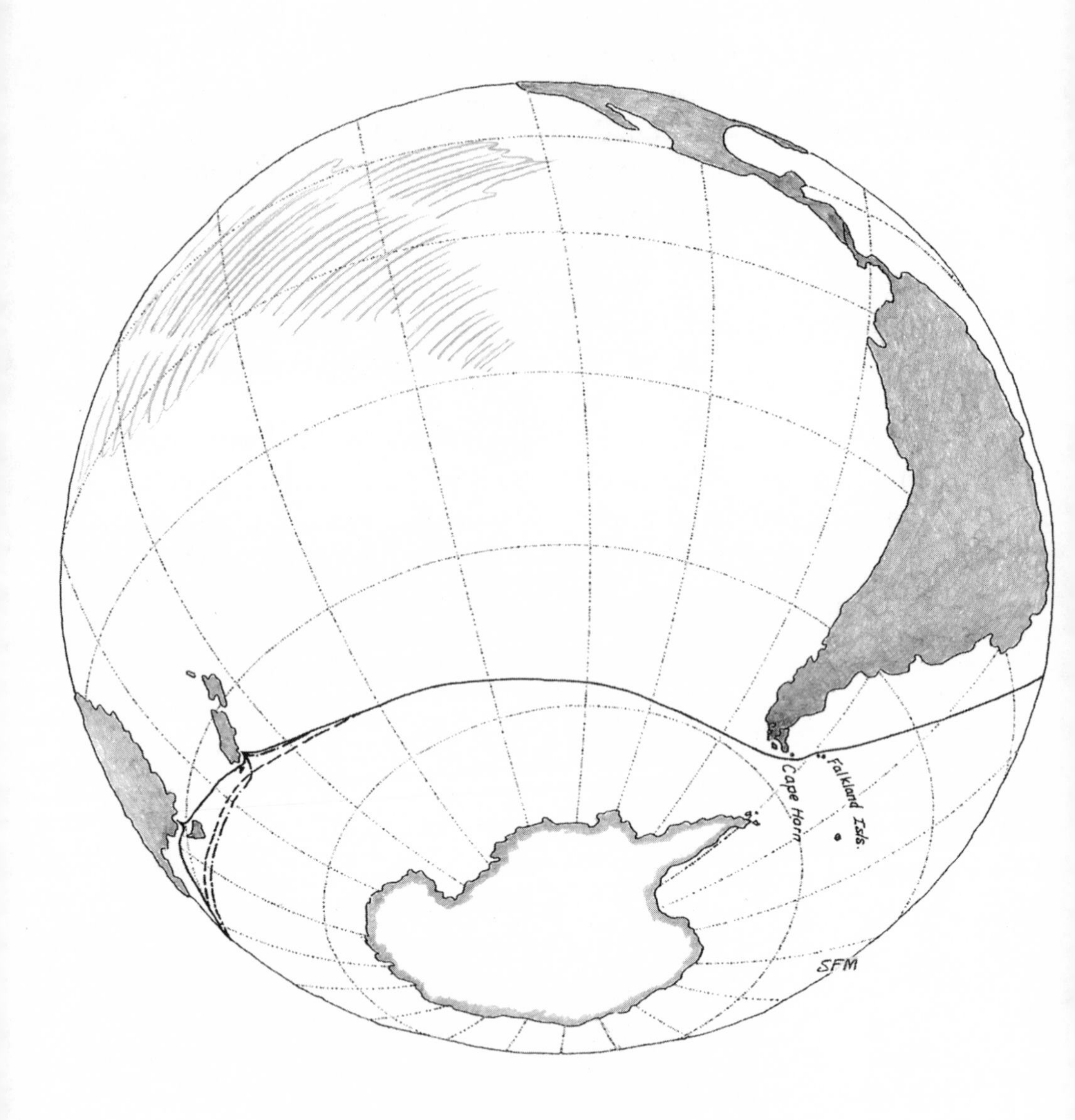

Sailing Route Across the Pacific Ocean

Cape Horn passages: Robin Knox-Johnston, 17 January, 1969
Bernard Moitessier, 5 February, 1969
Nigel Tetley, 18 March, 1969

S

23

BERNARD MOITESSIER CROSSED the Pacific fast. Daily runs: 146, 148, 143, 149, 148, 152, 166, 158, 147, 162, 169, 130, 111, 147, 142, 166. A thousand miles a week, week after week. The lightened *Joshua* was far fleeter than she had been three years earlier on the Tahiti-Alicante passage. Moitessier had evolved as a seaman. His daily runs through this same stretch of ocean were 20 to 40 miles greater than on the earlier voyage.

He had reduced Knox-Johnston's original 9-week lead to only 2½ weeks, and as the Englishman had suspected, he'd been luckier with the weather. He had the westerlies, but for many days past New Zealand they were light. However, Moitessier coaxed consistent runs out of *Joshua*. The midday temperature in his unheated cabin was in the 70s. He walked about the deck barefoot and was able to do his yoga exercises naked in the cockpit. He was more than halfway between New Zealand and the Horn before his first Pacific gale found him. It turned the sea white with foam, but it was not on a scale to trouble *Joshua* or elicit high-flown prose from her captain. The wind vane steered. Moitessier sat on his perch inside the hatch staring out of his turret at the fast-passing seas.

He ate well. Skinny all his life, with a tendency to lose weight, he began to gain, always for him a sign of a sympathetic environment. His unceasing close communion with the three constant physical elements of his world—his boat, the sea, and the weather around him—filled him with joy. And to complete the picture of happy asceticism, his hair and beard had grown long and matted until he resembled a sailing holy man.

Not since Captain Nemo had a man felt so comfortable and self-sufficient at sea. He had entered into a kind of seagoing stasis. The beginning and end of his long voyage grew remote, and deep in the vast middle he was untroubled by anything but the daily concerns of sailing ever onward. The rhythm of the sea, the endless passing of waves, the daily surging progress of *Joshua,* the now perfect vessel, and his own highly attenuated skills and sensations all blended into a harmonious chord that pealed loud and clear inside Moitessier and gave him peace. He was a man who had gone to the mountaintop and found the elusive thing he had been looking for, and he had grown reluctant to think much beyond the looming milestone of Cape Horn.

On January 21, after his first Pacific gale had passed, the course turned more southerly as Moitessier began his descent toward the Horn. As he dropped through higher latitudes, he left the fine weather behind. The sky grew overcast and the temperature fell. Waves broke over the boat and sluiced along the deck. Now he wore foul-weather clothing and seaboots outside. When he came below with soaking socks, he changed into dry ones to wear in the cabin and tried to discipline himself to put the wet socks back on before going topside to handle sails, but it was too easy not to. Socks and wet clothes now hung in the cabin, the hatches remained closed, and it grew damp below. Southern Ocean conditions prevailed.

He started wearing a safety harness, something he was not used to. Loïck Fougeron had given him one of the two harnesses that had come with *Captain Browne.* These were made of nylon webbing, and in the 1960s they were becoming popular on yachts, replacing the sailor's old standby, at extreme moments, of tying a rope around the waist. Once a harness comes aboard a boat, it seems foolhardy not to wear it, but it comes at a price. A harness interferes with one's natural and unconscious accommodation to a boat's movement, something acquired after a few days at sea. It makes one move awkwardly, stopping and starting to clip on and off every few feet. Harnesses have unquestionably saved people from going overboard, but they have also failed, come undone, broken, chafed through, and sent people to their deaths. An overreliance on them breeds an atrophy of the best of all devices to keep a sailor aboard: a fully developed horror of going overboard. A thoughtful, determined refusal to go overboard, a most careful premeditation and visualization of one's movements on deck, lines strung around the perimeter of a boat, and an overall design to ensure staying aboard put the occasional wearing of a safety harness in its proper place: an additional measure, rather than a single fallible device.

Moitessier's harness bore the name "Annie" marked with indelible ink. It had belonged to Annie Van de Wiele, an accomplished cruising voyager who, with her husband Louis, had sailed around the world in the steel yacht *Omoo* before they had built and sailed *Captain Browne* to the West Indies and sold her to Fougeron. Moitessier wore it mostly during the heaviest weather, when he clipped its snap-shackle to 3/16-inch steel wires stretched flat on deck, running from bow to stern on both sides of the boat. He learned the wire trick from Bill King, who used it aboard *Galway Blazer II,* which had no guardrails at the edge of its rounded decks. This allowed Moitessier to clip on once and move fore and aft while handling sail. He appreciated the harness and this wire system, but he also felt less mobile using it, and he felt the loss of his own more sensitive and certain link to his boat. He felt foolish ignoring the harness, however, so he tried to keep it with him—mostly in the pocket of his foul-weather jacket.

As *Joshua* drove farther south, the cold night sky made the stars and moon unusually luminous. Sometimes at sea, the moon's first appearance above a dark horizon can have a startling effect.

The points of its early or late crescents can resemble bright, unearthly horns coming out of the ocean a short distance away. It can suddenly pierce a clouded horizon and take on the characteristics of a city or an ocean liner. When one has been surprised by more than a few of these creepy apparitions, one reacts to weird night lights at sea by thinking, "Oh, yes, the moon doing one of its numbers again." So Moitessier thought one night close to Cape Horn when a slender spire of light rose from the sea into the clouds like a spotlight. How did the moon do that? he wondered. Then the "moonbeam" widened and glowed and played among the clouds until he realized the phenomenon he was seeing had nothing to do with the moon. He wondered, with a chill, if it was the "white arch" that Joshua Slocum had written about, the spectral white-squall forerunner of a great Cape Horn gale that had almost ended *Spray*'s voyage by carrying her into the "Milky Way," a cluster of reefs and white water at the edge of the Straits of Magellan.

Another spire rose, then another, and branches beamed out, spread, and soon covered half the southern sky, glowing and pulsating in pink and bluish tones like cold fire, and finally Moitessier knew that he was seeing the aurora australis, the southern sister of the northern lights. It lasted for nearly an hour, and he watched, dazzled, riven by its beauty.

Strong winds, but still no great gales pushed him on down toward the Horn. As he grew closer, he spent more and more time on deck, day and night, held by the sea, the heavens, the weather, his boat, taking it all in. He stood on *Joshua*'s cabin top, or on her plunging, twisting bow, holding onto the steel pulpit or the inner forestay, wild hair and beard streaming, staring ahead, upward, and at the sea around him, for hours, until driven below only by cold or hunger or exhaustion. These tended to, he returned to stand and watch again. The Ancient Mariner on a surfboard.

None but the callowest sailor can pass Cape Horn without gratitude and trembling, without being acutely aware of its history, of all the ships, seamen, and civilian passengers, men, women, and children wrecked, smashed, and drowned there. Deaths that often came at the end of weeks of the vilest discomfort and despair imaginable. For most sailors the Horn is their Everest; for Moitessier, it was Mount Olympus, a holy place, a crucible in which he, his seamanship, and the sea gods he believed in converged. For him it was Ultima Thule, to which he brought the eastern imagination of his childhood. It was going through the wardrobe to Narnia. He had passed it by once before, but after only 20 days at sea, and with his wife Françoise aboard to comfort and worry and distract him. Now after more than 5 months alone, he realized it was this totemic coordinate in time and place that he had been aiming for all along—not the race's end. As he flew toward it, he stood exposed and raw at the center of his world, sucking in every strand of sensation available to him.

He sailed through a very different sea than Knox-Johnston. There was not a whit of fancy in the Englishman's prosaic view of things, even when sighting the Horn. By comparison, Moitessier could have been on LSD. The beads of moisture on his sails were "living pearls"; phosphorescence in the waves became "globes of fire," reminding him that he had once tried harpooning such lights, thinking they were the eyes of giant squids.

> *Joshua* drives toward the Horn under the light of the stars and the somewhat distant tenderness of the moon. . . . I no longer know how far I have got, except that we long ago left the borders of too much behind.

On February 5 the strong winds finally increased to gale force, yet under a blue sky and a brilliant sun that turned the sea a deep violet. He had covered 171 miles in the previous 24 hours, and now drove on faster. Feeling no hunger, he ate nothing all day.

The Diego Ramirez islands appeared, a blue spot on the sunny horizon, in mid-afternoon. By twilight they were a speck far astern. The wind eased off with the coming night. Moitessier set his alarm for 0100 to be woken 20 miles from the Horn and turned in to sleep. When he woke, he knew from the angle of the risen moon through the portholes that he had slept through the alarm and possibly even passed the Horn. He went on deck and looked away to port, where he knew it lay. The night was clear, full of brilliant stars, and he saw only the moon with a cloud draping its moonshadow over the sea beneath it. The wind was lighter and had backed from the west to southwest, angling *Joshua*'s course 15 degrees closer toward the little island, the false cape.

Then the cloud moved, and there was Cape Horn beneath the moon, less than 10 miles away—a small, black, rocky shape against the starry sky at the edge of the dark sea. Moitessier was overcome with chills of euphoria.

The seaman's traditional rounding of Cape Horn was really the whole passage from 50 degrees south to 50 degrees south around the bottom of South America, either from the Pacific to the Atlantic, or from the Atlantic to the Pacific, the harder, meaner passage against the prevailing westerlies. That 1,000-mile passage contained so many attendant terrors—storms, drifting glacial ice, currents, and the screaming katabatic williwaws of Tierra del Fuego, Slocum's white-arched squall—that could stop a ship and shove her backward along her wake, making her lose in an afternoon sea miles that had taken weeks of desperate struggle to gain, that not until the latitude of 50 degrees south in the destination ocean had been reached could the Horn be safely said to be astern. That was the full meaning of rounding the false cape Moitessier saw across the moonlit sea.

He well knew it. He knew that so far he had been lucky. Not until he neared the Falkland Islands in three or four days' time would he be properly past Cape Horn.

Yet he celebrated the passing of the rock. The high-wire state of mind that had gripped him for two days now eased up. He went below, turned up the cabin light, made coffee and rolled a cigarette, and allowed his thoughts at last to turn toward his destination.

S

24

IN THE MIDDLE OF JANUARY, the *Sunday Times* photographed Françoise Moitessier, Clare Crowhurst, and Eve Tetley together aboard the *Discovery,* the ship that had carried British polar hero–bungler Captain Robert Scott on his first expedition to the Antarctic in 1901. The ship, moored on the Thames near the Tower of London, had tall masts that provided a suitably nautical frame for a photograph of the three sailors' wives. On Sunday, January 12, the newspaper ran an article with the macabre headline, "The Sea Widows They Left Behind."

Françoise Moitessier, who had already sailed around Cape Horn, declared that her own ambition was to be the first woman to sail alone around the world without touching land.

Clare Crowhurst provided a more prosaic, wifely point of view. She didn't have nightmares about her husband, she said, but her 7-year-old son, Roger, did, in which he saw his father standing at the door to his room, staring at him. Her 8-year-old boy, Simon, on the other hand, thought sailing around the world was nothing and planned to swim around when he was "old enough."

Eve Tetley was confident of her husband's chances. The

racers were only just past the halfway point, and since five of the nine starters had "gone down," she said, others were likely to drop out and his position would become stronger as the race went on.

The next morning her husband almost went down himself. At 0500 on Monday, January 13, Nigel Tetley was 450 miles south of Cape Leeuwin, western Australia, when a wave struck *Victress* along the whole length of her beam, with a single sledgehammer blow. She was lying ahull in a gale, drifting sideways at the time. Water tore through the wheelhouse curtains, through the cockpit doors, and into the cabin. Tetley, who was below, didn't see the wave coming. He only saw it smash against the cabin windows so hard he was sure they would break. Amazingly, they held. Two hours later the wind had risen to hurricane force, and the seas, Tetley wrote in his logbook, were unlike any he had seen before.

Waves or seas "unlike anything I'd ever seen" is an inadequate last-resort description, yet one frequently employed by hardened seamen in extreme conditions. Seasoned sailors come to know that their own impressions of great waves at sea, even when measured by eye against the known height of an object, such as the mast of one's boat, tend to be exaggerated by as much as 100 percent. More than 100 years of oceanographic studies, and now wave height measurements taken from satellite-sensitive GPS transponders on weather buoys, have shown that waves of 30 feet or higher are the rare product of unusually powerful winter storms in the high North Atlantic or the high latitudes of the Pacific. But even knowing this, what experienced sailors might rationally understand of the reality of waves at sea is driven from their minds and replaced by subjective terror. Fifteen-foot waves beneath a dark sky, driven by a shrieking wind, look terrible enough—great gray-green impersonal mountains with the density of concrete looming high overhead, and always more coming without pause or end. Being tossed about on them like flotsam removes every last vestige of a physical sense of security, so that the observer easily believes he or she may soon die in this nameless place far out at sea, far from shore and safety and loved ones. It is this dam-bursting of a lifetime of shored-up fears, reducing one to the most fearful childlike state, that makes the awful sea look at least twice as terrible as it really is.

The photographs sailors take of the great waves that impress them so at the height of a storm, are always later disappointing in their inability to convey what such a scene "felt like." Ironically, the impossible and wholly unrealistic computer-generated waves and conditions depicted in a film like *The Perfect Storm* do in fact provide very accurate impressions of what it *looks like* far out at sea in a terrible storm. It is their excessive exaggeration that mirrors the subjective impression of the human observer. Yet the movie feels safe. It comes without the horrifying realization that *this is real, there's no way out, nothing in all the world will save you now but luck*. This is what turns big waves into the vertiginous forms and shapes found only in nightmares.

Nigel Tetley, an experienced seaman, guessed that the great waves he saw that day in the Southern Ocean south of Australia were 80 feet high. He may have been right—maybe they *looked* 160 feet high. His writing is plain and not given to embellishment.

His great Southern Ocean storm continued all day. All day he believed that *Victress* would break up at any moment and he would be drowned. At the same time, the persistently hopeful side of human nature in him vowed that if he didn't die he would sail north to Albany and give up. But he didn't die, and the wind blew the trimaran away over the sea almost as fast as the onrushing waves, few of which hit her with any force after the first smash.

The next day the wind dropped to normal gale force, and when he saw how remarkably little damage had been done—the torn wheelhouse curtain, a soaked battery charger—he decided to carry on, if he could, as far as New Zealand, and see what things looked like then.

This was a starkly courageous decision. When *Suhaili* had finally lost her self-steering rudder a little farther east in the Great Australian Bight, Knox-Johnston had wavered and thought about giving up in Melbourne. It was the knowledge that he was in the lead, that he stood a chance of winning the race, that had pushed him on. Nigel Tetley had no such encouragement. Nor, having heard from his radio contacts of Moitessier's much faster progress, could he hope for the cash prize for the fastest voyage. He had only the cold comfort of knowing that these were the conditions he could expect now that he had left the Indian Ocean and sailed at last below the fortieth parallel. This was what the Southern Ocean was all about and he could look forward to three more months of it, with no chance of a prize at the end of it. Still he sailed east.

(Nearing western Australia a week earlier, he had contacted Perth Radio and found he had a call waiting for him. He was patched through to Dr. Francis Smith, president of the Western Australian Trimaraners Association. Dr. Smith sent him heartfelt greetings on behalf of all Australian trimaran owners. The safety of multihulls was then being seriously questioned in Australia; five such craft had been lost in Australian waters in the last year, fifteen dead from their crews. Authorities had demanded investigations. Nigel Tetley had appeared at a propitious moment, and found himself celebrated as the poster boy for the Aussie trimaran movement. His reluctance to let his brother multihullers down was probably a factor in his decision to keep going, along with his amazement at how well *Victress* had stood up to her first Southern Ocean drubbing.)

Most of the Golden Globe racers exhibited an abundance of Ulysses factor traits, but Tetley did not conform to the profile. He was a man with a steady job who one day simply read about a race and decided instantly to join it. He was often frightened and perhaps less sure of the reasons for his circumnavigation than the others, but once he had decided to go—very late and almost impulsively compared to the long-planned campaigns of his rivals—he stuck to his course. In his deceiving ordinariness, in the apparent wispiness of his motivation, and in the extraordinary steadiness of his resolve, he was the strangest of the nine.

Rough weather followed him across the Great Australian Bight. Like Moitessier, unlike Knox-Johnston, he passed to the south of Tasmania, a more direct route toward New Zealand. His sponsor, Music For Pleasure, hired an airplane to take pictures of *Victress* as she sailed south of Hobart, but the airplane couldn't find him. Tetley didn't wait around but headed across the Tasman Sea. Approaching New Zealand, he thought about sailing through Foveaux Strait, like Knox-Johnston, as a shortcut rather than going around Stewart Island, but overcast conditions gave him poor sextant readings and he wound up off the south side of Stewart Island. He cut between North and South Trap reefs and the southern tip of the island, then steered northeast for a day, close to the New Zealand coast, feasting on the green and rugged scenery that reminded him of the Scottish coastline he had seen from *Victress* on the Round Britain race. On February 2, he rounded Tairoa Head into Otago Harbour, where *Suhaili* had gone aground, and found a small fishing boat to take his package of mail and photographs. The fisherman offered him a crayfish, but the *Sunday Times* race rules, which would have termed such a gift "material assistance," forced Tetley to decline. His package was quickly forwarded to England, and the photograph of Tetley eating his lonely Christmas Day lunch appeared in the *Sunday Times* on February 9.

The New Zealand radio stations were forecasting a hurricane approaching from the north, so Tetley tacked back out of Otago Harbour and headed *Victress* out to sea. Hours later, at sunset, he had his last sight of green New Zealand, now a gray shape astern, disappearing into rain clouds. 4,700 miles of Southern Ocean lay between him and Cape Horn.

Donald Crowhurst's position was cloudier.

Newspapers reporting on the race could only give the same hazy locations that Crowhurst was sending to Rodney Hallworth, his sole media contact, who in turn issued his bullish Crowhurst bulletins to the press. On January 5, the *Sunday Times* stated that he was "reported" to be past Tristan da Cunha, which would "indicate" that he was sailing at over 1,000 miles per week.

He was, in fact, sailing slowly and desultorily off the Brazilian coast.

The *Sunday Times* published something about "its" race every week: this might be a half-page spread of the latest photographs it had received (Tetley at Christmas dinner, Moitessier practicing yoga on deck); or a breakdown of the latest positions, with headshots of the four sailors; or a learned essay by Sir Francis Chichester on the men's chances and the future of single-handed voyaging. Despite being advised by Captain Rich and Chichester that Crowhurst's positions were almost certainly impossible, the paper could hardly ignore the whereabouts of its fourth and last-place competitor; but it could not openly suggest skepticism. So it took the news it got from Hallworth, couched it inconclusively, and stuck it in a small paragraph at the end of its race reports.

Thus on Sunday, January 12, Crowhurst was reported to be in the Indian Ocean "by now." The following week he was "well into the Indian Ocean." The paper noted that his voyage's daily average was now up to 100 miles per day, and his expected date home was advanced to August 19.

Hallworth, with little or nothing to dress up his scanty information, plaintively cabled Crowhurst asking for weekly positions and mileage. On January 19, Crowhurst obliged with a reply, giving a position and weekly mileage as "100 southeast Gough 1086." (Gough is a small island south of Tristan da Cunha.) This was the transmission as written in Crowhurst's neatly maintained radio log. However, by the time it reached Hallworth, it read "100 southeast Tough." Hallworth, for whom the glass was always half full, took this to mean that Crowhurst was having a tough time southeast of Cape Town—1,200 miles farther east than Gough Island, and 4,000 miles from his true position.

In this cable Crowhurst also advised that he was sealing the cockpit floor hatch over his generator, and that future radio-transmissions, both radio-telephone and cable, would come much less frequently.

On the same day Crowhurst sent a cable to Stanley Best, mentioning damage to the boat's "skin," resulting in an "ill-found" boat. In few and ambiguous words, he implied he could only keep going toward the Horn, risking further damage, if Best would let him out of the clause in their contract that could require him to repay Best for the cost of the boat—forcing him, in other words, to buy back what could result in a worthless, half-wrecked boat if he pressed on at speed.

Then he closed down communication from the world. Nothing more was heard from him for 11 weeks.

His dizzyingly false positions reported in the press were now in excess of 4,000 miles away from his true location. It had become increasingly difficult to maintain and transmit a steady supply of false data, although he continued to record radio weather forecasts for areas far away, where he was supposed to be; he wrote these weather reports in painstaking detail in his radio log, often in triplicate as he received identical reports from other stations. But the emotional burden of this effort was proving too much for him, and he wanted to stop.

It has been suggested that Crowhurst initiated radio silence at this point to overcome the apparent obstacle of sending radio messages to stations in faraway Australia and New Zealand, as though he were in the Pacific, while all the time remaining in the Atlantic. But this would not have been a problem. On a Mercator-projected map of the world, his position off the coast of South America, roughly near Buenos Aires, does seem a long way from Australia or New Zealand. But on the true, round earth, he was no farther, as the signal flew across Antarctica, from Sydney (7,300 miles away) or Wellington (6,200 miles) than from Portishead Radio, north of London (7,000 miles), which he was able to reach without difficulty. Crowhurst knew this. He had simply lost the heart to make up positions to feed Rodney Hallworth and an eager press on a regular basis. Silence and presumption, after the steady apparent gains of the past few weeks, would now do a far better job of slipping him ever eastward than he could.

Rodney Hallworth provided additional grease. Crowhurst could not have had a better unwitting partner in deception than his zealously boostering publicist. With nothing but these two last cables to go on, Hallworth inferred a dramatic episode to pass on to the papers, and a perfect cover for prolonged radio silence: a huge wave had crashed over the stern of *Teignmouth Electron,* damaging the cockpit and stern of the boat. Repairs had required Crowhurst to drop sail for three days and seal off his generator compartment. In order to conserve his batteries, he would make only two more radio transmissions before arriving home.

"Crowhurst Limps On After Battering by Giant Wave," was the bold headline in the *Sunday Times* of January 26. The article said that Crowhurst was in serious trouble in the Indian Ocean, 700 miles east of the Cape of Good Hope. The following Sunday, February 2, he was "estimated" to be 1,300 miles east of Cape Town.

In that same February 2 issue of the *Sunday Times,* in an article about the race and the future of single-handed sailing, Sir Francis Chichester wrote, with great restraint, that there had been some "loose" claims for speeds and distances sailed, and he hoped that some sporting club would check and verify such claims. His cool skepticism, coming at the end of a long, windy, slightly pedantic article, was at variance with the more exciting race reports, and his lone voice went publicly ignored.

On board *Teignmouth Electron,* Crowhurst did have a real problem. The plywood skin of the starboard hull was indeed split, in several places, and a frame inside that hull had separated from the skin. The hull was leaking, and the trimaran could no longer be considered seaworthy. No great extremes of weather had been met with so far, so the damage, it would seem, was the result of poor construction, undoubtedly exacerbated by the rush of production spread between two different boatyards, which almost guaranteed problems. These might have been ironed out with the normal sea trials that any new boat needs to reveal and attend to problems. *Teignmouth Electron* was clearly far weaker and more vulnerable than Nigel Tetley's *Victress.*

Crowhurst now had the best possible and most honorable reason of all to give up. No shame would have come from putting into port with a damaged boat, but he was too deeply entrenched in deception now, and Rodney Hallworth's embellishment had given it a momentum of its own. Geographically he was too far from any port to limp into from his last supposed position. Thirteen hundred miles east of Cape Town would make Madagascar the most reasonable port to show up at, a long hike from the coast of Brazil. He could not give up now without exposing the whole sham.

Slowly he headed his boat for the South American coast.

Donald Crowhurst's route—

31 October 1968 — March 8, 1969

S

25

For most of January and February 1969, Crowhurst steered *Teignmouth Electron* in a slow zig-zag, moving roughly south off the coasts of Brazil and Uruguay. His daily mileages dropped to dawdling distances of 20 and 30 miles or less. His course was dictated more by which way the wind was blowing than a resolve to move in any direction. He had by this time abandoned any attempt at a circumnavigation.

But he was still at sea, and in a vessel that was taking on water. This presented him with a real problem that no deception or fantasy could push away.

One reason for his reduced mileage was that sailing the boat at any speed drove more water through the hull leaks. The longer this went unattended, the worse it would get. The split plywood skin, or sheathing, of the starboard hull was the sort of problem that might have been readily fixed at sea—Knox-Johnston's underwater caulking job was far trickier—if *Teignmouth Electron* had carried the tools, wood, screws, bolts, and general mix of chandlery and hardware that amount to basic and essential stores for any boat heading out to sea. Most of these supplies had been bought or provided for *Teignmouth Electron,* including spare pieces of plywood from Eastwoods yard in Norfolk, but in the confusion of his departure, it had all been left on the dock at Teignmouth—or unloaded from the boat—and Crowhurst had sailed without it. The cargo he had paid most attention to—the boxes and boxes of electrical and radio parts, the unfinished "computer," the dense spaghetti weave of wires going nowhere—were no help to a leaky boat. Crowhurst had overlooked two fundamental points reiterated over and over by the experts and sailors he most admired and read, Chichester and Eric Hiscock: (1) Keep it simple, and (2) Sailing and electronics are, in the long run, incompatible. Electric systems on boats are under constant assault by the marine environment, and failure is their most consistent condition (or certainly was in the 1960s).

Crowhurst's "revolutionary ketch" leaked and could not be patched. Its bilge pump was inoperable. *Teignmouth Electron* was a fool's ship, a fact Crowhurst knew all too well by now.

At the beginning of February, he began heading slowly toward land. As he went, he studied his Admiralty pilot book of sailing directions for South America, much as he had done when considering putting into Madeira. He read up on every small port, bay, and possible landing site along the Argentine coast. Due west of his present position was the heavily trafficked Rio de la Plata, the wide sea entrance to Buenos Aires, where a tattered trimaran would not go unnoticed. Much better was the quieter Golfo San Matías 600 miles to the south. The pilot book promised a small settlement, where he would surely find materials to repair the leaking hull, and a good anchorage. He made penciled marks in the book beside this section. But at 42 degrees south, Golfo San Matías was technically in the Southern Ocean. At his present 36 degrees south, Crowhurst was already experiencing strong weather; he could easily encounter ship-battering conditions in the week or more of sailing south into the forties that it would require to get there.

Eventually, proximity decided him. He closed with the coast at the wide bay of Bahia Samborombón, just south of the mouth of Rio de la Plata. The pilot book described an anchorage off a small river, Rio Salado, near the top of the bay. A group of sheds and buildings stood on the south bank of the river—a small nondescript place but with some signs of life. This sounded perfect. He made a list for himself of what he needed to find ashore: plywood, screws, and vindaloo paste for the curries he liked to make. He might as well have hoped for Marmite.

Crowhurst sailed into Bahia Samborombón on March 2. He saw lights ashore. Perhaps in indecision, he sailed out again, heading offshore for two days. Then he turned back into the bay. He lowered his sails off a resort town, Clemente de Tuyu, didn't like the look of it, raised sail again and headed farther up the bay to Rio Salado. He finally dropped anchor off Rio Salado at 0830 on March 6. There was deeper water around him, but Crowhurst had unwittingly anchored on a sandbank. The tide was ebbing; *Teignmouth Electron* was soon aground.

The trimaran was spotted by Nelson Messina, a 55-year-old fisherman who lived in a small house on the north bank of the river. He saw that the boat was aground and presumed its crew needed assistance. He set off to inform his neighbor, Santiago Franchessi, senior petty officer of the local Prefectura Nacional Marítima, whose post was one of the sheds mentioned by the pilot book: Crowhurst had run himself aground right under the nose of the local coast guard station. The job of Franchessi and his staff of two men and one dog was to observe the shipping entering and leaving Rio de la Plata. However, Rio Salado was a sleepy place with no telephone and only a dirt road winding up the coast to more prosperous parts. The coast guard station was a lowly, unimportant one.

Messina took Officer Franchessi and his junior recruit, Rubén Colli, out to the foreign yacht in his fishing boat. They were surprised to find only one man aboard. He had a patchy beard, and wore khakis and a red shirt. Franchessi greeted the sailor in Spanish; he responded in English, and they did not understand each other. The sailor gestured to the damaged starboard hull, and the point of his visit was immediately clear. Messina stepped aboard the yacht and tied his line to *Teignmouth Electron*'s mooring cleat. Then he stepped back aboard his own boat and pulled the trimaran into deeper water. According to the *Sunday Times* rules, this short tug off the sandbar knocked Crowhurst out of the race.

At 1100, they towed the yacht up the Rio Salado and moored it to the coast guard dock. Crowhurst noted the time and fact in his logbook.

The Prefectura's third man, Petty Officer Cristobal Dupuy, who had remained ashore manning the station, recorded the arrival of the yacht and the visitor's name in the station's logbook. He examined Crowhurst's passport, which gave the name of the bearer as Donald Charles Alfred Crowhurst. Dupuy dispensed with "Donald," which he took as an honorific similar to the Spanish *Don*. And he also left off the last name, which in a Spanish name is the unused matronymic, or mother's maiden name. "Charles Alfred of English nationality" was the name recorded in the Prefectura's log, and Donald Crowhurst didn't point out the mistake.

Señor Alfred spent half an hour trying to convey to the Argentinians the materials he wanted to repair his boat, but he could not make himself understood. He tried French, too, but this also was not understood, although Officer Franchessi recognized it as French. He took the foreigner outside to his Jeep and drove him 17 miles north along the coast road to Rancho Barreto, a former henhouse where Hector Salvati and his wife, Rose, and daughter, Marie, ran a small roadhouse restaurant. The Salvatis were French. Hector had once been a sergeant in the French army and had emigrated with his family to Argentina in 1950. Señor Alfred, it turned out, spoke excellent French.

He told them he was in a "regatta," that he had sailed from England on October 31, had rounded Cape Horn, and was on his way back to England, and would win the regatta if he could repair his boat.

Hector Salvati asked him how anyone would know he had sailed around Cape Horn, and the Englishman told him there was a machine on Cabo de Hornos that identified passing ships.

He then drew a map on a piece of wrapping paper showing them the sailing course around the world. He drew a second picture for the Salvatis, showing his trimaran, from the side, and from above. He wrote on this "Octobre 31–68." Then he drew a third picture, a rough map of the Atlantic, but showing a different route: England to a small island off South Africa, then to South America, then, in a lighter line, back to England—perhaps an attempt at confession. But neither the Salvatis nor Officer Franchessi understood the meaning of this last drawing, and the Englishman, talking excitedly and disconnectedly, wasn't making much sense.

Franchessi bought him a beer and explained that he had to use the Rancho's pay phone to call his superiors in La Plata and ask them what to do. Señor Alfred became disturbed, saying that if anybody knew he had stopped here, he would be disqualified from the regatta. But Franchessi made his call anyway, and the Englishman seemed to calm down.

His mood ashore was almost manic. He was very excitable, the Salvatis said much later, up and down in mood and laughing a great deal.

"Il faut vivre la vie," he said to Rose Salvati a number of times, laughing almost hysterically. "Life must be lived." She thought he was laughing at them, and she didn't believe his story. She thought he might be a smuggler.

Franchessi's superiors thought nothing of the stranded yacht. He was told to give the captain whatever he needed and let him sail when he was ready. Franchessi and the Englishman said good-bye to the Salvatis and returned to Rio Salado.

Crowhurst spent the night aboard his boat, tied to the coast guard dock. The next day, using materials given to him, he screwed two 18-inch-square pieces of plywood side by side over the splits in the starboard hull and painted them white. That evening, Petty Officer Dupuy and recruit Colli invited him to eat with them in the coast guard shed, where they lived. Franchessi, who was married, lived in a small house nearby. Crowhurst shaved for dinner, and the coast guard men fried him a steak of good Argentine beef and gave him wine. But the men couldn't understand each other, and they ate mostly in silence.

In the morning, Nelson Messina towed *Teignmouth Electron* back down the river to the sea. At 1400, the trimaran sailed away. Crowhurst had not found his vindaloo paste.

S

26

"ROUND THE WORLD! There is much in that sound to inspire proud feelings; but whereto does all that circumnavigation conduct? Only through numberless perils to the very point whence we started."

So mused Ishmael in *Moby-Dick*.

This was what Bernard Moitessier had been thinking. "Leaving from Plymouth and returning to Plymouth now seems like leaving from nowhere to go nowhere," he wrote after he rounded Cape Horn.

But he was eager to let friends and family know that he had safely passed the great rock. He sailed to the Falkland Islands, hoping to pass another packet of mail and photographs to a boat there, but by the time he closed with the islands, four days past the Horn and sleepless after another gale, he felt too exhausted to sail down the long narrow fjord into Port Stanley and back out to sea. He hove to for a time off Port Stanley Lighthouse, hoping to be seen, but he saw no sign of activity and turned away, heading north into the Atlantic.

But the red-hulled *Joshua* was sighted. Her position was reported to Lloyds in London and passed on to the *Sunday*

Times. This confirmed glimpse was fat meat for a starved press. With no word from Knox-Johnston in four months, Crowhurst's radio now silent, and only Tetley able to radiotelephone his occasional messages, the *Sunday Times* bulletins had been floundering for want of hard news about its race competitors. Here was real and exciting information at last: Moitessier was past the last great obstacle of the course and bound for home. His speed and position looked set to get him back to England first and fastest, a clean sweep of both prizes.

For the next few weeks his estimated positions—based on his consistent overall voyage mileage of 120 miles per day, run along his presumed track from the Falklands toward England—were confidently printed by the *Sunday Times.* On February 23, he was reported to be 1,250 miles east of the Plate River (Rio de la Plata). On March 2, the cunning Frenchman was 650 miles southeast of Trindade Island and was presumed, that weekend, to be crossing his outward track, making him the fastest nonstop circumnavigator, beating Chichester's record, and perhaps the first nonstop sailor around the world. "Moitessier on Last Stretch" was the paper's headine on March 9, predicting his arrival in Plymouth in just six weeks' time, on April 24.

France was readying itself for the homecoming of a national hero. Once Moitessier had accepted the Golden Globe and the £5,000 cash, an armada-sized fleet of French yachts and naval warships would escort *Joshua* back across the English Channel—or La Manche, as the French knew it—into home waters. Ashore, Moitessier would be awarded the Legion d'Honneur. He would eclipse Chichester, and even Tabarly, France's single-handed OSTAR hero. He would become the most famous yachtsman in the world.

And he knew it. That was the problem. Moitessier appreciated what his fame and books had brought him. More—much more—was coming, but he was afraid of its seductive hold on him. The yin and yang of his Asian-bred asceticism and his Westerner's worldly ego had always battled for supremacy in him, but seven months on his own at sea, his world stripped clean to the spare onward rush of his voyage, had tipped the scales heavily toward the unworldly, Buddhistic side of his nature.

"I am really fed up with false gods, always lying in wait, spiderlike, eating our liver, sucking our marrow," he wrote in his logbook.

Moitessier had been happiest in his early "vagabond" days at sea. It's those simpler times—the places, the people, the boats—he writes about most affectionately in his books that deal with his later, more famous exploits. He had found a measure of that simpler peace on his long, long voyage. He didn't want to give it up. He didn't want to lose himself forever in celebrity.

On Tuesday, March 18, Moitessier sailed into Cape Town Harbour, South Africa, 3,500 miles from where everyone thought he was, halfway up the Atlantic. It was almost exactly five months since *Joshua* had collided with the freighter *Orient Transporter* in Walker Bay, 50 miles southeast down the coast. Now, as the port captain's small launch circled *Joshua,* he heaved into a crewman's hands a 3-gallon plastic jerry can full of mail, his logbooks, rolls of still film, and ten reels of 16-millimeter movie film. He asked that it be delivered to his publisher, Robert Laffont at Editions Arthaud, Paris. Then he headed back out to sea. He passed close by a British Petroleum tanker, *British Argosy,* and slingshot a smaller can onto its deck, with a message to the *Sunday Times.*

> My intention is to continue the voyage, still nonstop, toward the Pacific Islands, where there is plenty of sun and more peace than in Europe. Please do not think I am trying to break a record. "Record" is a very stupid word at sea. I am continuing nonstop because I am happy at sea, and perhaps because I want to save my soul.

What he was saying was clear enough, but its meaning was almost too astounding to be grasped. It was repeated over and over as the story broke in the press: *Moitessier was not returning to England. He was abandoning the race, and his almost certain chance of being the double winner. He was sailing past South*

Africa for the second time and continuing on around the world.

Had he gone mad? the newspapers, sailors, people everywhere who had followed the race wondered. Françoise Moitessier, hearing the news out of the blue at the same time as everyone else, was shocked.

No, he wasn't mad, not by his lights. He had looked clearly into himself and seen the humbug of his ambition, the arbitrariness of the race's design. Plymouth meant nothing to him. Plymouth to Plymouth was going nowhere. Far better to go someplace he might enjoy. Better still if it was another 10,000-mile sail away and would extend his voyage. He was happy, and he wanted to stay that way.

A fresh southwesterly wind was blowing outside Cape Town Harbour. Moitessier hardened sheets and *Joshua* beat southeast toward the Cape of Good Hope. And the Indian Ocean, and beyond.

On that same day, Nigel Tetley rounded Cape Horn, turned the last corner, and headed with unequivocal desire for home.

He was more than ready to sail north out of the Southern Ocean. *Victress* had suffered two frighteningly violent episodes in the last few weeks, and the trimaran was no longer in any shape to please her Australian multihull boosters.

Three weeks earlier, on the evening of February 26, in not particularly large seas, a rogue wave (one much larger than those around it) reared up underneath *Victress*'s stern, hurtling her forward at tremendous speed on its crest. The steering cables broke (again) and she slewed sideways before the breaking crest. It smashed into the upper hull pushing it high into the air, until Tetley believed the boat tilted over on her side to an angle of 50 degrees, and he felt that she was about to cartwheel. Every movable object inside the cabin was hurled to the low side, and then the great wave boiled away beneath the trimaran and it slowly righted. It was the nearest he had come to a capsize, and it left him wondering, shakily, if he had simply been lucky so far or if this was a freak occurrence that would not happen again—not soon, at any rate. He couldn't know. He chose the latter explanation and pressed on.

On March 4, *Victress* was lying ahull, beam-on to seas, in storm conditions, and Tetley was inside the cabin, when another rogue wave struck the starboard side. This time the 6-foot-wide Perspex saloon window shattered before the weight of the solid sea that poured into the boat. At the same moment, the canvas curtain shielding the starboard side of the wheelhouse was blown away like tissue paper before a giant sneeze, and more water flooded into the wheelhouse and down into the cabin through the cockpit doors. The interior was awash with icy seawater.

Reading this, in a chair or tucked safely in bed, one tries to conjure the scene, the awfulness of what it must have been like. But the hours following such a catastrophic smashup inside a boat can scarcely be imagined. The sea that flooded the cabin, inundating everything, was perhaps 52°F. The air temperature outside *Victress* was around 48°F. The frigid water sloshed through the boat, drenching Tetley, mixing food, bedding, charts, books, clothes, his music tapes, kerosene, every large and tiny object into a swirling, lurching, icy stew. The gale that blew outside now blew into the former haven of the cabin through the cockpit doors and the 6-foot-wide smashed window. Tetley was as soaked through with seawater as if he'd jumped overboard. But there was no way to dry off, no dry clothes to change into, no way to get warm except by the frantic effort to save his life. Gone, utterly, was the thin membrane of shelter that permits the preposterous but precious and necessary illusion of security inside a boat, that sustains the warm life-force, bolstered by pictures of loved ones, books, music, the feasibility of making a hot meal, the unreasonable but persistent hope for one's chances, the barrier against despair and raw fear. All of that was swept away in an instant. At such a moment, the sailor fights for his life with all the desperation of a man in combat against a force he knows to be overwhelming. Still he struggles, and the struggle is what saves him. It gets him through the next few hours.

As so often happens in major swampings, the bilge pump was immediately clogged by the floating debris in the cabin. Tetley had to bail frantically with a bucket while hoping more waves didn't quickly pour in. As soon as he was able to, he found hammer and nails and boarded up the broken window opening with plywood. Then he tried to make some sense of the disorder below. Cleaning up from such a disaster appears at first impossible: everything is everywhere, only not where it belongs, so when you try to put something away, you first have to move what's in its place, and then find somewhere to put *that.* "A place for everything and everything in its place" is how it must be on a boat that contains a thousand discrete items, each crammed into its own special nook or cranny. Losing order can be like the nightmarish unraveling of an intricate puzzle.

Night came. Shivering uncontrollably with cold and shock, Tetley pulled his sodden sleeping bag around him and waited out the night, hoping every wave, which announced its approach with the hiss of its breaking crest, would not smash through his hasty repair and flood the cabin again.

In the morning he found *Victress* seriously damaged. The starboard hull had broken frames and a sprung deck, considerably weakening its structure, and the main cabin top had separated in places from the deck—the same thing that had happened to *Suhaili* during her first knockdown in the Southern Ocean west of Cape Town. The main hull was flexing and its longitudinal stringers were throwing off splinters as the wood wracked and twisted.

Tetley was amazed at *Victress*'s recovery from both of these hard knocks. He concluded that she was a fine seaboat, just not strong enough for the Southern Ocean. He wanted out. He decided to sail north to Valparaiso, put the boat up for sale, and fly home.

But Valparaiso, or any port, was still a long way away. He would be around the Horn in two weeks, and beyond lay the Atlantic and better weather. The next day, when the wind and waves had subsided and some order had been restored to the cabin, "sheer obstinacy" set in, and he headed east again. He pushed on hard, determined to get out of the Southern Ocean as fast as possible.

At 1400 on March 18, the sky cleared and Cape Horn and its island group lay ahead. As the afternoon wore on, the wind dropped and by evening Tetley was becalmed south of the fearsome rock. It was a welcome relief. He wasn't worried it would last. He lowered sail, had a celebratory dinner with a bottle of wine, and went to bed. He had accomplished an impressive sailing first: the first multihull to round Cape Horn. The Aussies would be thrilled.

A little over a week later, making a scheduled call to Robert Lindley, his radio contact at the *Sunday Times,* Tetley was stunned to learn of Bernard Moitessier's decision to drop out of the race. But he didn't for a moment think Moitessier had lost his mind. He knew for himself the whole story that lay behind such a decision. Moitessier had always said that such a voyage should not be considered a race; that all those who survived would be winners. He was saddened by the loss of his favorite competitor, but he thought it "very like Bernard," in whom he had detected no sense of rivalry. It is touching, reading Tetley's and Moitessier's accounts of the race, how often and fondly these two thought of each other during their long months at sea. Moitessier was constantly anxious about Tetley's safety in a multihull, writing at one point that he didn't know how he could take it if he learned that he would not hear from Nigel again. And Tetley thought of the Frenchman with unenvious admiration and happiness at his seamanship and fast passages. They had found much they liked in each other as sailors and men and formed a strong bond in their weeks together at Plymouth.

Tetley was, however, buoyed by the realization that with Moitessier out, he was now in the running for one or both prizes. Crowhurst was evidently making a very fast passage, but the last news from him was that he had sustained damage from a ferocious wave in the Indian Ocean. He could be out or pressing on at speed again. No one knew.

And there had been no word or sightings at all of Robin Knox-Johnston since he had sailed out of Otago Harbour, New Zealand, on November 21, four months before. Suddenly it seemed very possible that Nigel Tetley could be the first Golden Globe sailor to return to England and glory.

On Sunday, March 23, the same day newspapers reported Moitessier's dramatic change of mind, thirty British, American, and Portuguese vessels began a massive mid-Atlantic search for Robin Knox-Johnston. The ships were part of a NATO fleet already on exercises in the area. Planes from the U.S. Air Force base on the Azorean island of Terceira, which routinely made daily long-range patrols over Atlantic waters, also began looking for the small, battered ketch.

This followed weeks of mounting apprehension. "Fears Grow for Knox-Johnston" said many newspaper headlines. Even his sponsor, the *Sunday Mirror*, had somberly speculated in print whether he would ever be seen again. The location of the mid-Atlantic search was based on the supposition that if all had gone well and he had continued sailing at his voyage average of about 99 miles per day, he might now be nearing the Azores. But no one knew if he had rounded the Horn, or whether, after he had last been seen in New Zealand, he had even made it across the vast Pacific in his damaged boat, which was held together in places with string. Many yachting experts, including Sir Francis Chichester, thought it would be miraculous if Knox-Johnston had been able to keep going without putting in to some South American port for repairs.

Robin Knox-Johnston had kept to his 99-mile-per-day average, but *Suhaili* was too far away, about 1,000 miles southwest of the Azores, for the NATO search to find her.

He had sailed a lonely ocean, seeing no ships at all between New Zealand and Cape Horn. He was deeply aware of the anxiety his family would be suffering for him, and once round the Horn he planned to sail into Port Stanley to signal his position. But northeasterly winds blew him away as he approached the Falklands, and the specter of the Frenchman close astern stopped him from taking several days to beat in against the weather. He headed north, hoping instead to find a ship. In the South Atlantic he spotted a single freighter too far away to be seen or attract its attention. Once across the equator, however, his route began to intersect shipping lanes, and Knox-Johnston grew confident that he would soon be able to send word of his whereabouts back to England. He was in for a shock.

At last, on the night of March 10, a day after crossing the equator, he saw a ship coming toward him from the north. When it got close, he began signaling with his high-powered Aldis lamp, but there was no answer from the bridge. What the bloody hell was the officer on watch up to? Merchant Captain Knox-Johnston wondered. He lit a handheld flare and continued signaling. When there was still no response, he took the drastic measure of setting off a distress rocket flare. The whole sky around the ship was lit up in the sulfurous glow for 3 minutes as the flare drifted slowly down onto the sea in its small parachute. Knox-Johnston aimed the Aldis at the bridge again and finally there was a flickering answer. But as he started signaling his boat's name and identification numbers, the ship lost interest and steamed away. He set off another flare, and continued signaling until the ship disappeared over the horizon.

Knox-Johnston was outraged. He had given every proper indication of being a vessel in distress and had been ignored—worse, briefly acknowledged and then ignored. The nameless ship, too dark to identify, had also ignored a sacred tradition of the sea, backed by maritime law, that unless a vessel must put itself in danger to do so, it will *always* go to the assistance of another ship in distress.*

*This doesn't only apply to large ships; when my 27-foot yacht was sinking in the North Atlantic in 1983, a 900-foot-long American container ship, the *Almeria Lykes,* responded promptly to my Mayday

Knox-Johnston also had real cause to be alarmed at being so ignored. He had a stomach ache. At first he thought it was indigestion, but when the pain persisted and then moved into the side of his abdomen he feared it was appendicitis. He had foolishly overlooked antibiotics when taking aboard his medical supplies and had no way of checking such an infection. The nearest port was Belem, 1,000 miles to the west—ten days' sailing—though if he really did have appendicitis, it could kill him before he reached land. His only hope at sea would be help from a passing ship. The failure of this one to stop, despite distress flares, appalled him.

He was then crossing a shipping lane and saw a number of ships in the next few days, several of which came within half a mile of *Suhaili* (which by now, even without trying to attract attention, presented a desperate sight). None of them answered his signals. He was either not seen or seen and ignored. To an officer well and thoroughly trained in the venerable rules of the sea, this behavior was a "shattering revelation," Knox-Johnston wrote in his logbook. He came to the unhappy conclusion that the tradition of a fraternity of seamen watching out for one another on the high seas was no longer something that could be depended on, an unhappy thought for a single-hander.

He didn't have appendicitis. The "bully beef" had started to go off. He left this out of his diet for a while and got better—except for a recurrence on March 17, his thirtieth birthday, which he celebrated with a mixed grill.

On April 6, in light winds, he crossed another shipping lane, a busy one. All afternoon, ships passed him and ignored his constant efforts to signal them and attract their attention. Finally, late in the day, the British tanker *Mobil Acme* took proper notice of

call, left its course, and steamed up over the horizon to rescue me. No thought was given to the size of my boat or to the singular number of its crew. The *Almeria Lykes* was on its way from Rotterdam to Galveston, Texas. On its passage east across the Atlantic several weeks before, it had rescued a yachtsman who had suffered a heart attack and taken him to Bermuda.

him. Knox-Johnston and the officer on the bridge "spoke" to each other by Aldis lamp, with courtesy and efficient professionalism. Knox-Johnston gave *Suhaili*'s name and an ETA in Falmouth of two weeks. *Mobil Acme* responded with a "will-do" and "good luck." The ship radioed London immediately, and within two and a half hours of his sighting, Knox-Johnston's family received a phone call from Lloyds. It made the front page in every British Sunday paper.

What splendid news for England. The Frenchman was out; the three courageous Brits were now racing for home. Barring mishaps, Knox-Johnston now seemed certain to win the Golden Globe trophy for the first person to sail alone around the world without stopping. Nigel Tetley, heading north in the South Atlantic, looked good for the cash prize for fastest time, but Donald Crowhurst, the dark horse of the race, who "should now be approaching New Zealand," could still overtake him in elapsed speed.

At the other end of the earth, at the farthest reach of each sailor's due north, the British transarctic expedition, led by English explorer-author Wally Herbert, was at the same time approaching the North Pole after more than 400 days on the polar ice cap. For what? Another first in the annals of Arctic achievement: the first surface crossing of the Arctic Ocean by way of the North Pole. Never mind that it was not especially important or useful; it was wonderfully demanding. Herbert's team was traveling in the classic mode of men pulled by dogsled, evoking the glory years of England's heroic failures in polar exploration. It was another glorious, brutal, Ulyssean endeavor, and, pointless or otherwise, the island nation was showing that it still had, in spades, the stuff of which heroes are made.

Donald Crowhurst's route
March 10, 1969 — July 1, 1969

S

27

AFTER WAVING *adios* to Nelson Messina at Rio Salado, Donald Crowhurst headed north—toward England, of course, to win his regatta. But once over the horizon, beyond the eyes of Nelson Messina or the patrols of the Prefectura Nacional Marítima, he turned south. He was planning to rejoin the race.

His idea was to reappear at some point in April, break his radio silence, announce his position, and continue racing for home. Where, when, and how exactly he did this was crucial to his deception. He did not want to be spotted and identified by a ship before he was in the right position, and he had to work out his fake passage time across the Pacific and his ostensible date for rounding the Horn. He also hoped to get some 16-millimeter film footage of Roaring Forties conditions, and perhaps of the Falkland Islands. They might not be the Horn, but they were down there. It could be credibly easy, even commendably safe, to pass Cape Horn far enough offshore not to sight it: a glimpse of the Horn would be an almost irresistible temptation to a circumnavigator, but in thick weather even a minimally safe offing could leave the fabled rock hidden in stormwrack. A shot of the Falklands could vouch for a Southern Ocean passage to minds otherwise undisposed to serious doubts. Lastly, he wanted to send a radio message home, supposedly from the Pacific, via Wellington, New Zealand, before his arrival date at the Horn. This might be easier, he thought, if he got far enough south to bounce a signal over the lower, narrower spine of the Andes. So on March 10, he turned south from his position off Rio Salado and sailed down into the empty, loneliest region of the South Atlantic, away from the world's eyes, to plot his reemergence and strategy for taking the cash prize from Nigel Tetley.

His last purported position before initiating radio silence had been Gough Island, west of Cape Town, on January 15. From there to Cape Horn was approximately 13,000 miles. He decided to allow 90 days from Gough Island to Cape Horn, supposedly reaching there on April 15—an average of 144 miles per day; incredible but not quite impossible (Moitessier's monohull *Joshua* often did better). A reappearance shortly thereafter in the South Atlantic would put him—on elapsed time—comfortably ahead of Nigel Tetley. Then he would race for home as legitimately fast as he could.

He crossed the fortieth parallel around March 16 and continued zig-zagging south. He still had almost a month before reappearing more or less where he was. He soon ran into several days of Roaring Forties storm conditions, but by the time he was off the northern shores of the Falkland Islands on March 29, the weather had become unusually quiet. He shot some footage of a quiet sunset near Port Stanley. Then two more days of strong westerlies carried him away from the islands, northeast up the Atlantic.

Early in April, he began transmitting Morse cables on the frequency serving Wellington Radio, New Zealand, but with no success (not necessarily due to distance, but to atmospheric conditions). However, Radio General Pacheco in Buenos Aires picked him up on April 9 and repeatedly asked for his position. Typically, Crowhurst would not oblige with hard information. Instead, he used Radio General Pacheco to send his first words to England after eleven weeks of silence. It was a Morse message for Rodney Hallworth.

DEVON NEWS EXETER—HEADING DIGGER RAMREZ LOG KAPUT 17697 28TH WHAT'S NEW OCEANBASHINGWISE.

The cable was phoned to Hallworth as he was shaving on the morning of April 10. He immediately phoned Clare Crowhurst to give her the great news. Later, he sat down to work out what he could from this short, enigmatic message. How characteristic of Donald: the spare humor, the frustrating lack of detail and a pinpoint position. But this one was better than most: Crowhurst's log or logline had broken at 17,697 miles on March 28; "Digger Ramrez" was obviously the Diego Ramirez group, a scant 60 miles or so from Cape Horn. Why, Donald was sailing past the Horn as he read this! Hallworth realized. He prepared his press release.

Two days later, Sunday, April 13, half the London papers reported that Donald Crowhurst had rounded Cape Horn—almost a week before his intended date. The *Sunday Times* allowed that Crowhurst might have already rounded the Horn, but the paper also thought he could still have as much as 1,000 miles to go. At any rate, if he kept up his current speed, he could be back in Teignmouth sometime between June 24 and July 8. This would mean a circumnavigation of about 250 days—10 days faster than the current estimate for Tetley of 260 days. This would win Crowhurst the £5,000 cash prize.

Nobody seemed concerned that this message had come from Buenos Aires, the other side of the Andes from Crowhurst's supposed location somewhere in the Pacific. Radio contact at sea, as everyone knew by now, was a fluky business. Crowhurst's radio silence had been a month less than Knox-Johnston's, who had reappeared in the middle of the Atlantic. And Crowhurst was still the smaller story in the race. The papers were now full of news about Robin Knox-Johnston, "the surprising hero" in his small, Indian-built ketch, and his arrival any day now in Falmouth.

On board *Teignmouth Electron,* Crowhurst awaited Hallworth's reply with considerable anxiety. It was a month since he had sailed from Rio Salado. He'd had no news of the race since his last radio communications three months earlier. Now, having sent an ambiguous position, he waited to hear if he had been spotted at sea or reported ashore or if suspicions at home had been aroused and the game was up. The reply, 3 days later, contained no hint of trouble.

> YOU'RE ONLY TWO WEEKS BEHIND TETLEY PHOTO FINISH WILL MAKE GREAT NEWS STOP ROBIN DUE ONE TO TWO WEEKS—RODNEY.

Encouraged, and suddenly freed from his isolation chamber, Crowhurst became relatively garrulous on the radio. He sent back cables describing the smell of wood smoke on the wind off the Falkland Islands. (Hallworth heard poetry in the spare lines; *Wood Smoke on the Wind* should be the title of Crowhurst's book, he thought.) In another, he peevishly quibbled with the term "race-winner" when applied to Knox-Johnston, suggesting an even distinction between first home and fastest time.

In frequent if not regular communication with both General Pacheco Radio in Buenos Aires, and now Portishead Radio in England, he learned of Moitessier's abandonment of the race and Tetley's position, supposedly far ahead of him. (Bizarrely, the tracks of the two trimarans came so close together on March 24 that the two men might actually have passed within sight of each other, Crowhurst headed south toward the Falklands, Tetley past them going north. The weather at the time was stormy and visibility would have been poor—but what a surprise that would have been for both men! And perhaps a much different ending.)

Heading north, now "past" Cape Horn, Crowhurst's fake and true positions merged, and in the middle of April he began to race home for real.

Nigel Tetley felt him coming. He had heard that both Knox-Johnston and Crowhurst had reappeared after months of silence and doubt, and were alive and well and bashing on. Like

Moitessier, Tetley felt the rarest of kinship for his fellow racers, like soldiers facing a common enemy, and he was always happy and relieved to hear that they were alive, safe, and doing well.

But he was now very aware that the prizes that had so recently seemed within his grasp were slipping away. He would not be first home, and it looked as if Crowhurst, on elapsed time, might overtake him. He would be out of the money, effectively in last place.

So he pushed on harder than ever. *Victress* was now so damaged as to be a write-off by the time he reached England. He accepted that; her last and only remaining function now was to get him there. The worst seas and weather were behind her. She had only the gentle tropics and spring in the North Atlantic to face. His job was not to spare her, nor even, any longer, to save her—only to keep her afloat long enough to complete the voyage while moving as fast as possible.

But what had begun as the slow, steady, normal attrition of breakage and failure—bits of molding coming off, windows leaking, decks and outer hulls leaking—had been accelerated and made worse by the hard months and hammering damage of the Southern Ocean. The self-steering rudder had vanished into the deep shortly past the Horn; Tetley didn't have a spare, but by then he had become, like Knox-Johnston, so adept at balancing the boat with sails under a variety of conditions that this was a slight loss, more of a nuisance. The wire cable between the wheel and the main rudder broke again, but he replaced the wire as he had before. The fiberglass sheathing had begun to delaminate from the side of the port hull, long flaps of it had peeled away revealing the bare wood beneath. The plywood had no seams, like those in a rounded, conventionally planked hull, only a few joints where sections of plywood met over the frames; but without the watertight barrier of the fiberglass, seawater began to seep through those joints and undermine the strength of the glue and fastenings there. At this moment, when *Victress* needed all the sparing and gentleness he could give her, Tetley found himself forced to drive her on harder than ever if he was to keep ahead of the relentless gaining of Donald Crowhurst.

She couldn't take it. One morning at dawn, he found that a large section of deck between the main and port hulls was disintegrating. The beams beneath the deck, supporting it, were splintered and broken. More importantly than supporting the deck, these beams were part of the once rigid cross-arm structure keeping the port hull attached to the main hull.

Just as bad, the plywood in the bow of the port hull was holed and split, frames were broken and separated from the hull skin, and water was filling the hull. When he climbed down into the port hull and squatted knee-deep in water, he could see the deck lifting overhead, the frames broken away from the hull sides, and the plywood skin flexing in and out with each wave. The trimaran was coming apart, literally separating into its components.

It appeared that Tetley's race was over. A bitter blow, and all the more so since he was at that point only 60 miles short of crossing his outward track and circling the world. But *Victress* was not actually sinking—not yet. Tetley thought he could patch the bow and stiffen the hull sufficiently to allow him to make Recife, 200 miles away on the Brazilian coast.

Like all his fellow competitors (except Crowhurst), Tetley had loaded *Victress* with a supply of wood, tools, nuts, bolts, odd pieces of metal—that well-found boat's portable hardware store—and as he began to patch the port hull, he pondered the larger damage. He rooted through his materials, found what he needed, and the job stretched across 2 days—2 days in which to think of how far he had come, how close he was to home. When he was finished, the structure between the port and main hulls was strengthened with jury-rigged cross beams. The worst of the holes in the port hull were patched, but so much water was still coming in that he drilled holes below the waterline in the port bow's forward compartment to let water out—the water inside now stayed at sea level.

The repair was rough, but if he could now sail on toward Brazil, Tetley decided, he could bloody well sail toward England.

And he would tie the knot. On the evening of April 22, at

6°50' South, 30°38' West, *Victress* reached a place on the featureless ocean where she had been exactly six months earlier and crossed the track of her outward-bound route. Nigel Tetley had sailed around the world, the first person to do so in a multihull. But coordinates of latitude and longitude are abstractions, and when they represent nothing but stretches of seawater they don't make good newspaper and television pictures. Port-to-port is the proper framework for such a voyage, the tidy start-and-finish that record books, newspaper rules, and a hero-worshipping public all recognize. Tetley was 4,200 miles from Plymouth—5,000 as the sailor sails, following a course for the fairest winds. His voyage would come to its proper end in another month and a half, if luck and *Victress* held up.

28

ALONE AMONG HIS COMPETITORS, the "distressingly normal" Robin Knox-Johnston had never doubted himself. Blinkers at the periphery of his imagination had precluded any glimpse of inner failings, equivocation, or uncertainty that might have given him pause or made him give up. He had worried only about his cuts and bruises, his eye, the symmetry of his mustache, his water supply, and whether *Suhaili* and her gear would last the distance. He and the boat had endured, and 1,000 miles from England, after 29,000 sailed, these worries began to recede.

Early on the morning of April 7, he saw land for the first time since passing Cape Horn two and a half months earlier on January 17. A smudge appeared on the southeast horizon; it was Flores and Corvo, the small northwestern outriggers of the Azores.

Five days later, on Saturday, April 12, another ship, *Mungo* of Le Havre, came up astern and began to overtake *Suhaili*. Knox-Johnston—still unable to raise anybody on the radio—had no confirmation that *Mobil Acme* had passed on his signals, so he began signaling to *Mungo*, but the ship appeared, as usual, to ignore him and steamed past. Five minutes later he looked up and saw that she had turned around and was coming back. Soon,

Mungo began signaling *him*. When Knox-Johnston sent *Suhaili*'s name, the crew on the bridge began to wave—they knew him. Moments later, talking with *Mungo*'s radio operator over the radio (on a short-range frequency), he heard the astonishing news that Moitessier, whose red ketch he had long feared he might see coming up over the horizon astern of him any time, was actually in the Indian Ocean heading around the world a second time. He could hardly believe it, hardly allow himself the rush of relief.

The radio operator clearly knew all about the race, and when Knox-Johnston confirmed the name of his boat, he said, "Yes, that's right." He agreed to pass on a message to the *Sunday Mirror*. The two men chatted for a while and then *Mungo* steamed away. But the news was out. In the middle of the afternoon another French ship, a tanker, *Marriotte,* steamed close by and gave *Suhaili* three blasts of its horn.

Knox-Johnston's solipsistic shell was broken at last. For five months—since saying good-bye to the crayfishermen at Otago Harbour, New Zealand—he'd had no contact with the remembered world beyond the horizon, no tangible proof that anyone knew of him or cared what he was up to with this insane business of slogging across oceans at a walking pace. Now ships came out of their way to hoot at him, and he had regained a connection with the world.

The next evening—that Sunday, April 13, when many newspapers were reporting that Donald Crowhurst had rounded Cape Horn—Knox-Johnston tried calling the General Post Office high-frequency station at Baldock, Hertfordshire, north of London. Months of attempting to contact the station had brought no result, but that night he magically got through. After chatting for a few minutes, the Baldock operator asked him if he wanted to speak to anyone on the phone, and Knox-Johnston gave his parents' number. His younger brother Mike picked up the phone and "nearly went through the roof." Now he knew they could stop worrying about him, and after hearing that everyone at home was well, he could stop worrying about them.

Mike confirmed what he'd learned from *Mungo,* that Moitessier was still heading east in the Southern Ocean—Knox-Johnston had half-wondered if this was true or whether the French radio operator had fed him a story so that he might relax his effort. He heard that the two trimarans were in the South Atlantic, Nigel Tetley off the coast of Brazil, and Donald Crowhurst past the Horn and pressing hard behind him, racing each other for the elapsed time prize. Mike also told him about the boats carrying the press and his family that would be coming out to meet him when he neared land.

Suddenly it all seemed real. After ten months of a journey whose end had always stretched incomprehensibly far ahead in time and distance, Knox-Johnston allowed the full realization and thrill of what he had done, and what was coming, to wash over him. Unless anything very unusual now happened, it looked as though he would reach England in about a week's time, the first person to sail around the world alone, nonstop. He got out the whiskey bottle, went on deck, and poured a dram for *Suhaili* over her stern. He poured another into the sea for Shony, an ancient English god of the sea. Then he raised the bottle to his lips.

With the isolation of months now past him, ships and radio stations seeing and hearing him, Knox-Johnston suddenly found himself feeling like a sailor again, a land-linked fellow between ports, instead of a boundlessly pelagic creature. And very shortly afterward, like a fox chased by hounds. He was now in regular radio contact with the *Sunday Mirror,* passing them his positions and ETAs (constantly revised as weather conditions affected his progress) in code, so the *Mirror* could get its boat to him first and scoop its rivals with the first photographs of the returning hero.

At noon on Friday, April 18, *Suhaili* was 280 miles from Falmouth. Knox-Johnston was closing with the western approaches, the sea area west of Britain and France where shipping lanes bearing traffic from all points of the Atlantic converge into a narrowing stream. That night he was surrounded by the lights of many ships and had to remain awake in the cockpit, flares at hand, ready to take evasive action if any other vessel came too close.

Soon after midnight (Saturday morning) a well-lit ship that had been overtaking *Suhaili* slowed and took up station half a mile astern. Another smaller vessel approached, slowed, and fell in with the first. Knox-Johnston signaled the larger one with his Aldis lamp, requesting its name. *Queen of the Isles,* the ship answered. It was one of the vessels his brother Mike had said would be meeting him. It moved in close, cameras flashing along its deck in the dark. Knox-Johnston's mother and father were aboard and the three of them shouted hellos across the dark water. The smaller boat was *Fathomer,* a former rescue launch chartered by the *Sunday Mirror* and carrying the *Suhaili* Supporters Club—Knox-Johnston's crew that had sailed with him from London to Falmouth ten months earlier: the *Sunday Mirror* reporter and photographer covering his story and his editor from the publishing firm of Cassell.

Queen of the Isles and *Fathomer* then stood off and kept pace with *Suhaili* through the remaining hours of darkness, alerting other shipping while Knox-Johnston went below and got some sleep.

When he woke in the morning he found the two escorts gone. The weather had deteriorated and a gale was blowing from the southeast, pushing *Suhaili* off course to the north. Frustrated but unable to make much headway in the right direction, he hove to. During the afternoon *Queen of the Isles* and *Fathomer* reappeared. Now he was able to see his parents for the first time in 310 days, and wave and shout to the Supporters Club. And they were able to see him: *Suhaili* was streaked with rust, her paint chipped, hull sprouting seaweed along the waterline; and the returning hero was bearded, unwashed, and wearing filthy oilskins, but immensely cheerful.

Then the wind eased sufficiently to allow Knox-Johnston to raise sail and the little fleet proceeded toward England at the speed of its smallest member.

"Round-World Robin Battles to the End" was the front page, front-and-center headline in the *Sunday Times* the next morning, beneath a photograph of the bearded sailor aboard his battered ketch. The front-page article described his struggle so near home against a rising gale. The paper noted that if Bernard Moitessier had not given up the race, he might easily have been nearing Plymouth at the same moment.

Inside, on page 2, was a profile of "Solo Robin: The Surprising Hero." The boy who built a canoe while on holiday; the student who failed A-level physics but turned into a "brilliant" navigator; the stubborn Merchant Marine captain who had soaked up generations of sea lore. It was written by Murray Sayle, the *Sunday Times* reporter who had favored Tahiti Bill Howell for sponsorship over the unknown Knox-Johnston, who might have been a surprising hero to Sayle and the yachting experts who had dismissed his chances a year earlier but not to those who knew him. Again and again, his friends and family described him in the same stolid terms: "He is the sort of bloke who does what he sets out to do."

The full force of the stubborness that had taken him around the world against everything the wind and seas could throw at him was needed to carry him through the last few hundred miles. The weather was not cooperating with homecoming plans. *Suhaili,* as doughty a sea boat as a hero could want, was a poor performer to windward in the strong conditions of England's spring weather and the racing channel tides. At the center of a growing armada of well-wishers' yachts and boats chartered by the press, Knox-Johnston tacked slowly toward Cornwall all Sunday and Monday. The wind abated Monday night, but at 0900 on Tuesday morning, April 22, when he was 6 miles from Pendennis Point at the entrance to Falmouth Harbour, the wind rose again to gale force, blowing directly off the shore he was trying to reach, and pushed *Suhaili* away to the east. All morning and through the afternoon, Knox-Johnston beat the last few miles into Falmouth.

At 1525 he sailed between Black Rock and Pendennis Point at the harbor entrance. A cannon fired. People on boats and ashore cheered. The surprising hero in his rough Indian boat had won the great race.

Shortly afterward, when Knox-Johnston had smoothed his

water in the lee of land, Her Majesty's Customs and Excise launch ranged alongside *Suhaili.* Officers leapt aboard the battered ketch to perform their mandatory duty.

"Where from?" asked the senior officer, struggling with his composure.

"Falmouth," said Robin Knox-Johnston.

S

29

NIGEL TETLEY DROVE *Victress* north through the northeast trades and on into the belt of variable winds around the North Atlantic's midocean ridge of high pressure, sometimes given the attractive-sounding name, the "Bermuda high." But for sailors, this is no vacation spot: it is a place of light and fluky winds from all directions, where progress slows and frustration mounts.

So it was for Tetley, always aware of Donald Crowhurst in the other, apparently much faster trimaran behind him; and his tantalizing closeness to home and the end of his long voyage. Through the early part of May he moved from the western to the eastern half of the North Atlantic, climbing slowly north through the latitudes of the Cape Verde and Canary Islands, often having to beat into the northerly winds that streamed down the eastern edge of the Bermuda high, like morning fog drifting around the base of a mesa.

In the third week of May, as he reached the Azores, the wind strengthened. A fresh northwesterly on the beam pushed *Victress* fast through the 60-mile-wide channel between São Miguel and Terceira, a welcome burst of speed in the right direction. Tetley hoped the wind would hold. But on Tuesday, May 20—four weeks to the day after Knox-Johnston had reached Falmouth—the wind strengthened and reached force 7 (about 30 knots). A heavy sea built up on the port side and jolted the boat with each wave. Worried about the boat's weakened state, Tetley reefed the sails, but *Victress* sailed on almost as fast. It was just a short summer blow, he believed, looking at his May pilot chart, which showed no winds above force 7 for the sea area he was sailing through. But as the afternoon wore on, the wind rose steadily to force 9, a strong gale. In her present condition, he didn't want to run the trimaran through heavy seas in the dark, so at nightfall he lowered all sail and *Victress* lay ahull, drifting quietly away before the wind as she had so often before in much severer weather. Tetley turned in to catch some sleep.

He woke at midnight. He heard a scraping, wracking sound in the bow. Knowing his boat and its every piece so well, the sound conveyed a picture to his mind of what had happened: the damaged bow of the port hull had broken away and was caught between the hulls. He had already drilled holes in the bow, so it didn't really matter if it fell off; the watertight bulkhead just aft of the bow would keep water out of the rest of the hull, but the opened hull would certainly slow him down. He had to get up and clear away the wreckage. All this he thought in a moment. Then he reached for the light switch.

Water was pouring into the main hull. This he hadn't pictured. Dashing up on deck, he saw the port bow was gone, as he had thought, but in going it had smashed a hole in the bow of the main hull. When he returned to the cabin it was flooding at a rate that told him instinctively not to bother locating the leak—quantity told him everything. He reached for the radio and broadcast every sailor's most awful message.

There is a form to it. Somehow one already knows it by heart. *Mayday Mayday Mayday. This is the sailing yacht* Victress *at latitude thirty-nine degrees ten minutes north, longitude twenty-four degrees thirty minutes west. I am sinking and I request assistance. Mayday Mayday Mayday . . .*

An immediate, crisp, professional reply from a Dutch vessel

allayed panic. But by the time he had finished speaking with the ship's radio operator, water was swirling around his legs. It was time to get out the life raft. Still in the dark of night, he dragged it on deck, tied off its painter, and heaved it overboard where it began automatically to inflate. Into the raft he threw the film and logbooks that contained the record of his voyage; his sextant, chronometer, camera, binoculars, and a handheld radio transmitter; and some warm clothes. By then water had filled half the main cabin and the boat's movement was sluggish and dead.

As he climbed into the raft, some piece of the trimaran snagged its sea-anchor line and held it close.

"Give over, Vicky, I have to leave you!" Tetley yelled. He found the line holding him to his sinking boat, cut it, and soon drifted clear.

The waves were still high and the life raft went spinning and dipping away like an amusement park ride. Lights shone aboard *Victress* a moment longer, but then the sea closed over the batteries and they suddenly went out, and Tetley lost sight of her in the moonless dark.

He had no time for feelings just then. His predicament drove thoughts of *Victress* from his mind, and he busied himself trying to get the small emergency transmitter working. He continued sending his Mayday through the dark night but got no response.

When daylight came, he was able to read the transmitter's instructions and saw that he hadn't rigged the set's aerial. Once he'd done that and started transmitting again, he immediately made contact with an American rescue plane alerted by the Dutch ship and already searching for him. By midday, the Hercules aircraft from the U.S. Air Force 57th Rescue Squadron based in the Azores was circling overhead. At 1740 that afternoon he was picked out of the sea by the nearest available ship, an Italian tanker, *Pampero*, on charter to British Petroleum. They were happy to rescue him.

After eight months on his own, Tetley found that once he

started, he could not stop talking; he chattered compulsively to *Pampero*'s captain and crew. It kept him from brooding over the loss of *Victress,* except at night when he was alone in the small cabin they gave him. Then his mind flew back to his voyage, and the agonizing closeness of its completion. *Victress* had sunk a bare thousand miles from England.

Eight days later, *Pampero* docked at Trinidad in the West Indies. Eve had flown out to meet her husband and was there when he arrived.

It was over, Tetley thought. But he was to find that the Golden Globe race threw a long shadow.

S

30

NEWS OF NIGEL TETLEY'S SINKING reached Donald Crowhurst two days later, on May 23, in a cable from Clare. He was now the only remaining competitor in the race.

If no accident or mishap disabled *Teignmouth Electron* before reaching England, Crowhurst would be the winner of the *Sunday Times'* £5,000 cash prize. He would join Robin Knox-Johnston, Nigel Tetley, Sir Francis Chichester, and the other *Sunday Times* judges and experts for the Golden Globe celebratory dinner aboard the tall ship *Cutty Sark,* where they would swap stories about their trials in the Southern Ocean and confirm their standing in the small company of men that has been called the Cape Horn breed. Of these, Knox-Johnston, Tetley, and Crowhurst would be placed among the rarest of the elite, the solo Cape Horners.

It was the sort of glory Crowhurst had always yearned for. The notoriety and speed of his voyage would turn his company Electron Utilisation into a solid success. Book and merchandising deals would be forthcoming. The brilliance and superiority of Donald Crowhurst would be acknowledged by the world.

Also, Captain Craig Rich of the London Institute of Navigation, Sir Francis Chichester, and others would examine his logbooks and navigation records.

In fact, Chichester was already drafting a letter to Robert Riddell, the *Sunday Times* race secretary, asking for details of Crowhurst's messages and position statements, particularly his last message before leaving the South Atlantic and entering the Southern Ocean near the Cape of Good Hope and his next message about nearing the Horn ("Digger Ramrez"). "We need to know why the silence from the Cape to the Horn (from an electronics engineer too). . . . Why did he never give exact positions? It also appeared that he had an extraordinary increase of speed on entering the S. Ocean; I think he claimed 13,000 miles in 10 weeks, or something, which seems most peculiar considering his slow speed for the previous long passage to the Cape, and the succeeding 8,000 miles (Horn-home)." *Claimed.* Chichester had put the numbers and his own sea sense together and the conclusion was, to him, inescapable.

Crowhurst already felt the weight of scrutiny that awaited him. It was one thing to make up a story in the lonely, solipsistic space of *Teignmouth Electron*'s cramped cabin and feed it to the ecstatically credulous, geographically ignorant Rodney Hallworth, who passed it on to an equally gullible and wanton press. It was quite another to lay the lie before a committee of sea dogs and savants who had really done what he had only guessed at and pretended. Crowhurst knew this; he was a highly intelligent man. But he had chosen not to dwell on it. Now, only a few weeks away from stepping ashore into a klieg light of illumination and surrendering his logbooks, the fullest ramifications of his deception swept over him.

Crowhurst began to coast. He delayed, he zig-zagged, he let the wind blow the boat where it would. In the preceding weeks, since breaking radio silence, he had sailed faster and more steadily than almost any other period of his voyage, even clocking a genuine 200-mile-plus run in the 24 hours between noon of May 4 and May 5. But from May 23—the day he learned of Tetley's sinking—onward, his progress up the Atlantic became erratic. He passed out of the steady southeast trade winds and entered the Doldrums, the hot, steamy, thundery band of stagnant air and light and fluky winds either side of the equator. *Teignmouth Electron* ghosted through the water while Crowhurst, naked and streaming sweat, sat in the messy overheated cabin amid the detritus of his grand plan—wires snaking to nowhere, radios, boxes of spare parts, and a contradictory set of logbooks—trying to see his way home and clear.

Early in June, his Marconi transmitter failed. Suddenly his newfound voice, which he had been exercising since ending his self-imposed radio silence, the precious link to the world outside the claustrophobic cabin, beyond the empty horizon, was taken from him. For Crowhurst, the breakdown of his core electronic device was unhingeing. For the next two weeks *Teignmouth Electron* drifted slowly north, largely untended, while he devoted all his efforts to fixing the transmitter. He spent 16 hours a day sitting in the boiling cabin, surrounded by the cannibalized innards of radios and open tins of food, while he soldered and tinkered with wires and transistors, ate when he remembered to, lost in his work, fascinated, challenged, sustained by the one realm he truly understood.

The sea—the watery blue reality beyond the cabin, the discipline of seamanship, the purpose of his adventure—receded.

In the cooler, dark, early hours of June 22, Crowhurst fixed his radio and finally made Morse contact with Portishead Radio. He immediately sent cables to his wife Clare and to Rodney Hallworth.

Then, as the sun rose, the cabin temperature increased, and so did the heat coming from the repaired radio. For much of the rest of that day, Crowhurst sat hunched beside it, exchanging cables with Hallworth, who was already working on deals and syndication rights, and with Donald Kerr of the BBC, who wanted to arrange a rendezvous for boats and helicopters to meet Crowhurst offshore. The welcome, the clamor, the end of the voyage, the end of the game, loomed.

On Tuesday, June 24, Donald Crowhurst turned away from it all. He turned away from the world and plunged deep into himself.

At the top of a clean page in his logbook—following weeks

of comment-free mathematical workings of his celestial sights—he wrote a title: "Philosophy."

He began by discussing Einstein, whose book, *Relativity: The Special and General Theory,* was one of the few he had brought along on his voyage to read. Einstein had written the book to explain his theory to a general audience; in its day it was as well-known and as widely unread as Stephen Hawking's later explanation of the universe, *A Brief History of Time*. But for Crowhurst, reading it over and over in the isolation chamber of *Teignmouth Electron*'s lonely cabin, Einstein's statements took on the truth and gravity of holy writ.

One paragraph made a profound impact on him:

> That light requires the same time to traverse the path A to M as for the path B to M is in reality neither a supposition nor a hypothesis about the physical nature of light, but a *stipulation* which I can make of my own free will in order to arrive at a definition of simultaneity.

Einstein was only stating, or appropriating, a definition of the word "simultaneity" for the purpose of his argument. But to Crowhurst this Einsteinian exercise of free will appeared to be a godlike control of physics, of the universe. "You can't do that!" wrote Crowhurst, imagining a dialogue between himself and Einstein. "Nevertheless I have just *done* it," answered Albert. Crowhurst did not doubt Einstein's authority to take such control. He took it as an example of the power of a superior mind. This led him deep into a maze of tortured logic.

He was soon writing this:

> I introduce this idea $\sqrt{-1}$ because [it] leads directly to the dark tunnel of the space-time continuum, and once technology emerges from this tunnel the "world" will "end" (I believe about the year 2000, as often prophesied) in the sense that we will have access to the means of "extraphysical" existence, making the need for physical existence superfluous.

As he wrote, Crowhurst was listening to the radio. Beside his philosophical writing, he now made annotations of what he was hearing: "1430 gmt, 24th, Radio Volna Europa. 1435: Hysterical laughter."

He continued writing. Through the day, into the night, all through the next day.

At 1700 on June 25, when he had been writing for about thirty hours, a Norwegian cargo ship, *Cuyahoga,* passed close by *Teignmouth Electron.* Crowhurst appeared on deck and waved cheerfully as the ship steamed by. The *Cuyahoga*'s captain wrote in his logbook that the man on the trimaran had a beard, wore khaki shorts, and appeared to be in good shape. Crowhurst had spent the day writing a history of the past 2,000 years, with a further look back to the time of cavemen, illustrating the way exceptional men have, through the shock of their genius, changed society through the ages. At some point in this history, he put down his pencil, climbed up on deck, and waved at the *Cuyahoga.*

Over the next week, for 8 days from Tuesday, June 24, to midday Tuesday, July 1, Crowhurst wrote 25,000 words in his logbook (equivalent to almost a third of this book), stopping only to eat or nap as need overtook him. His hand flew across the pages, bearing down hard, the urgency of what he had to say outstripping the need to sharpen his pencil. His neat engineer's handwriting now grew large and irregular, the strokes thick with emphasis. The pages became dense with notes crawling around the margins, circled and crammed between distant paragraphs, as insight upon insight struck him and deepened his revelations. He wrote in a white heat of possession.

Stuff like this:

> The arrival of each parasite brings about an increase in the tempo of the Drama, causing first-order differentials in its own lifetime within the host, and second-order differentials within the host to the host, etc, etc . . .

> And yet, and yet—*if* creative abstraction is to act as a vehicle for the new entity, and to leave its hitherto stable state it lies within the power of creative abstraction to produce the phenomenon!!!!!!!!!!!!!! We can bring it about by creative abstraction!

> Now we must be very careful about getting the answer right. We are at the point where our powers of abstraction are powerful enough to do tremendous damage. . . . Like nuclear chain reactions in the matter system, our whole system of creative abstraction can be brought to the point of "take off." . . . By writing these words I do signal for the process to begin. . . .

> Mathematicians and engineers used to the techniques of system analysis will skim through my complete work in less than an hour. At the end of that time problems that have beset humanity for thousands of years will have been solved for them.

Despair and the moral burden of deception had lifted and been replaced by the exhilaration of seeing a great truth. "I feel like somebody who's been given a tremendous opportunity to impart a message—some profound observation that will save the world," he'd confided to his tape recorder months before. It was something he had always wanted, believing himself cleverer than the normal run of men, and ready for a chance to prove it. Now that message had been delivered to him in the peculiarly receptive vessel that he had made for himself, and he was in a fever to write it down and pass it on to the world.

Was it another fake? A pose? A few pages of such writing could be made up by anyone with an ear for the the ravings of psychotic breakdown. Novelists and screenwriters do it all the time, sometimes convincingly. Nothing, however, but a genuinely deranged mind could spend 150 consecutive hours producing 25,000 words of such passionately insane verbiage.

There was, however, a consistent theme to Crowhurst's psychosis: that in the end, by an act of will, a person of superintelligence—a great mathematician—could alter, and deliver himself from, the bonds and rules and obligations of the physical world.

Crowhurst went from a functioning, if cheating, competitor, a sender of rational cables, to the total abandonment of navigation and boathandling, and deep into scribbling madness, in the space of a few days. But it had been long in coming, since his earliest days at sea when he had faced the "bloody awful decision." From that point on, when he made the most rational and sane appraisal of his impossible situation, he had seen no way to go forward, yet no way to retreat. It was, at its root, a moral dilemma, and there his reason had foundered. Crowhurst had the cleverness, possibly, but not the conscience to carry off his hoax.

When Crowhurst looked up a week later, he had no idea what time it was. Nor even what day. His Hamilton chronometer and both of his (prequartz) wristwatches had run down and stopped. His last navigational entry had been on June 23.

He went up on deck. It was daylight, but he saw the moon—full—just above the horizon. He went below and opened his nautical almanac, which contained tables giving the phase of the moon and times, of its rise and set in Greenwich mean time. He concluded that it must now be June 30. He worked out the time to be 0410 GMT. He added an hour to make this British standard time to conform with the times of BBC broadcasts. It was 0510, then, on June 30. He wrote the time and date in his logbook, with the note that he was starting the clocks again.

Then he realized he must be mistaken: if it was 5 in the morning English time, it would still be dark where he was, 40 degrees of longitude west of Greenwich, in the middle of the Atlantic Ocean. He had made the stupidest mistake.

He wrote in his logbook:

June 30 5 10 MAX POSS ERROR

After studying the nautical almanac again, he decided it must now be July 1. And as close as he could gauge, it was 10 in the morning, British standard time. As a navigator, for whom time, accurate to the second, is of religious importance—the navigator's life literally depends upon it—he had slipped up badly. Now he watched every second, for time was ticking to a countdown.

He wrote:

EXACT POS July 1 10 03

It was a position in time. He had no need of a geographic position. He was past all that.

The minutes and seconds ticked by. Twenty minutes later, he wrote:

10 23 40 Cannot see any "purpose" in game.

10 29 No game man can devise
is harmless. The truth is that there
can only be one chess master. . . .

there can only be one perfect beauty
that is the great beauty of truth.
No man may do more than all
that he is capable of doing. The perfect
way is the way of reconciliation
Once there is a possibility of reconciliation
there may not a need for making
errors. Now is revealed the true
nature and purpose and power
of the game my offence I am
I am what I am and I
see the nature of my offence

I will only resign this game
if you will agree that

the next occasion that this
game is played it will be played
according to the
rules that are devised by
my great god who has
revealed at last to his son
not only the exact nature
of the reason for games but
has also revealed the truth of
the way of the ending of the
next game that
 It is finished—
 It is finished
IT IS THE MERCY

Against a great truth, the petty rules and structure of his voyage now seemed to Donald Crowhurst, as they had to Bernard Moitessier, irrelevant. Now that the truth had been revealed to him, and he had written it down for the world to find, his voyage was over. Finished.

The minutes and seconds had got away from him. He recorded them again:

11 15 00 It is the end of my
 my game the truth
 has been revealed and it will
 be done as my family require me
 to do it

11 17 00 It is the time for your
 move to begin

 I have not need to prolong
 the game

 It has been a good game that
 must be ended at the

I will play this game when
I choose I will resign the
game 11 20 40 There is
no reason for harmful

He had reached the bottom of the logbook page. There was no more room to write, and time was ticking along.

Time might, if he didn't watch it, even get away from him. So he unscrewed his round brass chronometer from the bulkhead and took it with him.

31

IT WAS 0750 IN THE MORNING of July 10 when the Royal Mail vessel *Picardy*'s officer on watch spotted the yacht. The weather was fine, the wind light, the sea almost flat, but the boat had only its mizzen sail up and appeared to be drifting aimlessly. The ship and the yacht converged at 33°11' N, 40°28' W—roughly in the middle of the Atlantic, halfway along the shipping lane between Europe and the Caribbean; about 600 miles southwest of the Azores or 1,800 miles southwest of England. The officer on watch called the captain to the bridge.

Captain Richard Box ordered his ship slowed and its course altered to pass close by the drifting yacht. It was a trimaran. The name, *Teignmouth Electron,* was clearly visible, painted on the stern and bow of its main hull. No one was on deck. Its crew was keeping a poor lookout for the *Picardy* to be able to get so close without being seen. Box ordered the foghorn sounded—that would get them out of bed. But the three loud blasts in the still morning produced no activity aboard the yacht. Now Captain Box was concerned; the yacht's crew might be ill or incapacitated. He ordered the engines stopped.

Chief Officer Joseph Clark and three crewmen were lowered overside in the ship's boat. They motored over the calm water to the yacht, and Clark climbed aboard. He stepped below into the cabin.

He found a squalid scene. Dirty dishes were piled in the sink; three large radio transmitter/receivers lay on the table and shelves, their insides exposed, wires, components scattered everywhere. A filthy sleeping bag lay on the bunk in the narrow forward cabin.

Clark found three blue foolscap books—two logbooks and a radio log—stacked neatly on the chart table. Beside them, carefully arranged, lay two navigation plotting sheets with positions worked out on them. Clark flipped through the logbooks. They were both full of writing. In one he found the last navigation entry; it was dated June 23, two weeks earlier.

Back on deck, Clark saw that the life raft was still lashed in place. The situation appeared tragic, but not mysterious: a lone yachtsman had fallen overboard—but not as a result of bad weather. Sitting next to the radios on the cabin table, a soldering iron still lay balanced on a milk tin: no unexpected wave had toppled the skipper off the deck.

Aboard the *Picardy,* a crewman remembered the yacht's name. He produced a clipping from the London *Sunday Times*. There was a drawing of the same yacht in an article about the Golden Globe race: *Teignmouth Electron,* a trimaran; Donald Crowhurst was its captain, from Bridgwater, Somerset.

Captain Box sent a cable to his ship's owners in London, describing the situation. Lloyds was notified. The U.S. Air Force, which had found Nigel Tetley, began an air search. *Teignmouth Electron* was hoisted aboard the *Picardy,* which then began a search of the surrounding ocean.

That evening, two Bridgwater policemen drove to the Crowhursts' house with the news. Clare took the children upstairs. They sat on the bed and she told them that the boat had been found and their father wasn't on it. But there would be a search, and he would be found. Then she began to cry.

Soon other cars, driven by reporters, began arriving. Clare Crowhurst would make no statement other than to say that she knew her husband was alive.

The next day the search for Donald Crowhurst was called off.

Two days later, Sunday, July 13, London's newspapers were full of the tragedy. The *Sunday Times* launched the Donald Crowhurst Appeal Fund for Crowhurst's widow and children. Robin Knox-Johnston, now the default winner of the £5,000 cash prize for fastest time, donated the money to the fund. The *Sunday Times* kicked in another £5,000. Mr. Arthur Bladon, chairman of the Teignmouth Finance and General Purposes Committee, declared he would recommend the launch of a local fund. The BBC announced it would donate the fees it would have paid Crowhurst for the film he was shooting on his voyage. The Royal Mail Line said it would return the trimaran to England at its own expense. Stanley Best waived his claim to the boat in favor of the Crowhurst fund.

The *Sunday Times* suggested that a guardrail, something *Teignmouth Electron* did not have, might have saved Crowhurst from falling overboard. There was speculation that he had not worn a harness (film taken by Crowhurst aboard—in fair weather—shows him not wearing one). Sir Francis Chichester always wore one, the *Sunday Times* noted. But Robin Knox-Johnston was quoted saying that he seldom wore a harness, finding that it hampered his movement around the boat. But even without a harness, he thought it unlikely that Crowhurst could come so far and simply fall overboard. He believed the only explanation could be "some dreadful accident." The paper noted that weather reports indicated calm conditions had prevailed at the time of Crowhurst's disappearance in the area where his boat had been found. Quotes from Clare Crowhurst and Rodney Hallworth summed up Donald Crowhurst as an adventurer who lived life to the fullest and followed his dream. *Il faut vivre la vie.*

And there were these sentiments from Chichester: "It is very sad that such an extraordinary accident should have occurred to such a gallant sailor after such a memorable voyage and so near home. But before he was lost, he had accomplished something near to his heart, having circumnavigated the world."

Rodney Hallworth, together with Nicholas Tomalin and Frank Herrmann, a reporter and photographer from the *Sunday Times,* flew to Santo Domingo in the Dominican Republic to await the *Picardy.* As their Boeing 707 flew over the water where Crowhurst had disappeared, Hallworth asked the group to observe a moment's silence.

When the *Picardy* docked, Captain Box took Hallworth alone to his cabin. He had read enough of the logbooks to know what had happened, and he urged Hallworth to rip out the "philosophy" pages for the sake of Crowhurst's family. Hallworth reflexively complied. But the next day, when the reporters read through the navigational logbooks and it became clear that Crowhurst had never left the Atlantic, Hallworth showed them the pages he was holding. They returned to England and conferred with the editors at the *Sunday Times.*

It was a sensational story, but not the one the newspaper had wanted. The *Sunday Times* had made it all too easy for anyone, unexamined, untried, unknown, to join its race and, together with Rodney Hallworth, had been an unwitting but eager partner in Crowhurst's great deception. The true story was also painful salt in the wound already being suffered by Clare Crowhurst and her children. But there could be no stopping it.

Donald Crowhurst's deception, madness, and presumed suicide were front-page news in all the national British papers on Sunday, July 27. On its front page, the *Sunday Times* published a sober statement about its decision to release the full story, in part necessitated by the existence of its Crowhurst Appeal Fund, which it intended to continue to support.

In the same statement, Sir Francis Chichester now abruptly changed his tune and publicly revealed his private doubts: "As chairman of the judges of the Golden Globe race I had decided some time ago that Donald Crowhurst's log must be scrutinised as soon as possible."

Inside the paper was reporter Nicholas Tomalin's account of the tragic voyage of Donald Crowhurst.

Months later, the Teignmouth Finance and General Purposes Committee officially commended Rodney Hallworth for "the terrific publicity reaped from the Donald Crowhurst saga." Arthur Bladon, the committee's chairman, estimated that the Devon town had gained £1,500,000 of free national and international publicity. "We have had this extremely cheaply," Mr. Bladon told the committee, "and I hope the town appreciates it."

S

32

STRANGE HOW AN ARTICLE about a race around the world, found by chance in his Sunday newspaper at the foot of his bunk, deflected the trajectory of Nigel Tetley's life and sent it spinning away with unstoppable momentum. It was not the sea that continued to hold him in its grip long after he arrived home, but a sea change that would not give him up to his old life and loved ones ashore. He had driven himself as hard as *Victress*. He had come so very far—too far—to have lost so close to the end and to let it go. The race was over, but he found there was no going back to the way things had been before.

He had wondered, often, why he was sailing around the world. But by the time of his sinking, he had found enough reason in the act of the voyage itself, and the tidy geometry of completion. A circular shape, fused without a break, had formed in his mind, and the compulsion remained to express it.

He was awarded a £1,000 consolation prize by the *Sunday Times*. He put it toward the building of a new trimaran. He planned to enter the new boat in the 1972 OSTAR and then head around the world alone again, trying for a new fastest circumnavigation record.

The boat was built by sailor and boatbuilder Derek Kelsall,

at Sandwich Marina, Kent. Tetley had met Kelsall several years earlier, when they had both sailed their trimarans in the Round-Britain Race, which Kelsall's yacht, *Toria,* had won. They became good friends.

"Nigel's circumnavigation was never recognized for what it was," Kelsall recalled later, "a truly remarkable effort in a most unsuitable craft. I can think of nothing that was right about that boat for that race. The outcome could have been so different if he had not been pushing hard to beat Crowhurst back to the UK."

The new boat, *Miss Vicky,* a 60-foot trimaran, was finished and completed sea trials by the end of 1971. Tetley and Eve moved aboard and moored it on the River Stour at Sandwich.

Now retired from the navy, Tetley wrote a book about his part in the Golden Globe race. It was full of the decency and generosity he had felt for his fellow racers, but it was too modest and told little of his inner voyage. It was about as dull as a boating book can be, and it sold poorly.

After the cost of the new boat, Tetley didn't have any money left for the sails, food, and equipment needed for his new transatlantic and round-the-world efforts, so he again looked for sponsors.

Kelsall saw Tetley daily at this time: "Nigel came to my office most days to get his mail, and he often talked about the latest potential sponsor letter he was waiting for. The one thing that Nigel did, that did not make sense to me, was that he would write to one potential sponsor at a time and then wait for that reply. He seemed to put all his faith in the last company he approached, but it was one letter only. I told Nigel that a couple of years before, Geoffrey Williams had successfully got *Sir Thomas Lipton* sponsored (for the 1968 OSTAR), but Geoff had written 2,000 letters."

Perhaps Tetley sent prospective backers his book, for he met with uniform rejection.

Derek Kelsall last saw Nigel Tetley on Wednesday, February 2, 1972. "He was his usual pleasant self the last day he collected mail, and I believe I was the last known contact before he went missing. There had never been any indication of a problem other than the search for funds. Everything was dependent on the search for a sponsor."

Three days later, on Saturday, February 5, Tetley was found hanging from a tree in Ewell Minnis Woods, near Dover.

Most people are horrified by the specter of great waves and storms at sea. "Aren't you afraid?" they ask sailors again and again. The truth is that such physical dangers are readily coped with; as conditions worsen, there is much to do aboard a boat in peril at sea. Even if disaster is the final result, the steps taken to avert it are clear at the time, and keep a sailor busy. One may be afraid, but action is a blessing that usually allays the deepest fears and doubts, and once a few storms are weathered, one acquires a comforting faith in one's efforts. After surviving the seas and terrors of Cape Horn and the Southern Ocean, Nigel Tetley's greatest danger rose up inside him, inescapably close. It found him at home, on dry land, in the company of friends and loved ones, where most people do not fear to go.

"Apart from his wife Eve, I probably knew Nigel better than anyone else at the time of his suicide," said Derek Kelsall. "That is not to say that I knew him well. Perhaps no one did. I don't believe there ever was a reasonable explanation of the suicide. There were a few stories, as there are with most boat people."

S

Epilogue

After ten months at sea, Bernard Moitessier finally dropped anchor at Papeete, Tahiti, on Saturday, June 21, 1969. He wrote his story of the race, *The Long Way*, which was a best-seller in France, and has remained in print in French and English for the last thirty years. It will doubtless continue so as long as people read books about the sea. He remained mostly in Polynesia, with occasional voyages to the United States and New Zealand. He and Françoise did not stay together. Almost everybody he ever met loved Bernard Moitessier, and he loved them all back, freely. Being his woman was a tough role. There would be two more. He had a child, Stephan, by one of them.

In 1980, he sailed *Joshua* to San Francisco, where he remained for two years. In 1982, he sailed south, headed back to the South Pacific. With him, just for the first leg, was actor Klaus Kinski, who was thinking at the time about sailing around the world. Moitessier dropped him off in Cabo San Lucas, Mexico. He stayed 12 hours too long. Before he could raise his anchor and beat out to sea, a now-famous storm tore into the fleet of yachts anchored off Cabo San Lucas and threw most of them—including *Joshua*—onto the beach, then buried them with sand. Moitessier couldn't face the damage done to *Joshua*, and he gave the boat to friends. It was his third shipwreck, and each had propelled him into a new phase of his life, each time with a new boat. So it was again. Other friends built him a new steel cutter, *Tamata,* in Point Richmond on San Francisco Bay. In 1983 he sailed to Hawaii, and from there back to Tahiti.

Moitessier died of cancer in France on June 16, 1994. He is buried in the Breton town of Le Bono, in a graveyard filled with seamen.

In 1990, the restored *Joshua* was acquired by the maritime museum of La Rochelle, France, where she now sails as part of a cruising school.

John Ridgway started an adventure school in Ardmore, Scotland. He sailed a maxi yacht, *English Rose VI,* around the world with a crew in the Whitbread Round-the-World-Race and circumnavigated a second time with his wife and children.

Chay Blyth remained fascinated with the masochistic aspects guaranteed by a single-handed circumnavigation. He persuaded the British Steel Corporation to finance a new boat, a 50-footer christened *British Steel* (made, naturally, of steel), which he then sailed successfully alone and nonstop around the world in 1970–1971—but the "wrong way." He sailed west-about, into the teeth of the westerlies of the Roaring Forties, which, despite Robin Knox-Johnston's few weeks of frustration, are the prevailing winds of the Southern Ocean. It was a brutal voyage, and Blyth seems to have thoroughly enjoyed it. He has remained a prominent figure in British yachting circles, a patron and participant of long voyages characterized by their hardship.

Bill King repaired *Galway Blazer II* and set out again in 1969 for a nonstop circumnavigation. Problems forced him to give up at Gibraltar, but he remained gripped by the adventure. He tried once more, in 1971. Off Australia, *Galway Blazer* was struck with great force and holed by what King later believed was a great white shark. Stuffing the hole with sails, King made it to Fremantle, Australia, where *Galway Blazer* was repaired. He set off some time later, completing his circumnavigation, by way of Cape Horn, in 1973.

Loïck Fougeron and Alex Carozzo retreated from the public eye.

At the end of the Golden Globe race, the *Sunday Mirror* sent Robin Knox-Johnston to visit the psychiatrist who had pronounced him "distressingly normal" before his voyage. The misdiagnosis was once again confirmed.

Robin Knox-Johnston has made a life of being England's preeminent yachtsman. He has become rich and famous, and has been showered with honorary degrees and every imaginable maritime award. In 1994, with New Zealand's preeminent yachtsman, Peter Blake, he sailed nonstop around the world again, this time in a giant catamaran. Their 74-day, 22-hour circumnavigation was the fastest record—until Olivier de Kersauson, a Frenchman, shortened it by three days.

Knox-Johnston continued to sail *Suhaili,* voyaging among other places to the icy seas of eastern Greenland with England's preeminent mountaineer Chris Bonington. In 1995, in his middle age, he was knighted by the queen for services to sailing. By then, with his beard gray, the sea years showing in his face, favored by his sovereign, Sir Robin Knox-Johnston had come to resemble exactly the Elizabethan sea heroes of his youth who had watched over him on his epochal voyage.

After thirty-five years of hard service, Knox-Johnston donated *Suhaili* to the National Maritime Museum at Greenwich, where she now lives, enshrined in glory, on permanent exhibition in a glass-roofed museum gallery called Neptune Court. She is heeled slightly on a 40-foot block of blue plastic waves, sails raised but slack, no one at her helm. Noble as such a berth may be, it is living death for a wooden boat. Out of her natural element, *Suhaili*'s planks are drying out and shrinking, her seams are opening up, the long cracks along her hull indicating the onset of decay. She is passing into history.

The Royal Mail Line retreated from its initial good intentions to transport *Teignmouth Electron* back to England. The trimaran was sold cheaply at auction in Jamaica to a man named Bunnie Francis, who used it to take tourists out day-sailing in Montego Bay. Sometimes he sailed with a calypso band aboard. But an increase in crime hit the tourist business, and Bunnie Francis sold the boat to a Canadian diver, Winston McDermott, for $12,000. McDermott had read about the Golden Globe race and knew what the boat had been through. It was a curiosity for him, but he also planned to use the trimaran for his scuba-diving business on Grand Cayman Island. McDermott and a young Jamaican he employed to sleep aboard the boat and look after it believed it was haunted. They said they heard footsteps walking around on deck.

One night, the trimaran was damaged in a hurricane on the island of Cayman Brac. McDermott hired a crane to haul it out of the water to make repairs, but he never got around to doing the work. He moved to Florida and the trimaran remained high and dry on Cayman Brac.

It's still there, lying in the weeds near the shore, heeled over on two hulls, like a strange carcass, sun-bleached and forgotten. The name *Teignmouth Electron* is still just visible, in faded paint, on the bow and stern of the main hull. Over the years, people have unbolted bits of it and stripped it of any useful piece of gear. All that remains now, other than its empty plywood hulls and deck, are its galley sink and pieces of its toilet, which lie on the ground between the hulls.

Inside the main hull are tangles of old wire, going nowhere.

In 1999, British artist Tacita Dean, who had become interested in Donald Crowhurst's story, was invited by the National Maritime Museum to exhibit photographs she had taken at Cayman Brac of the abandoned *Teignmouth Electron*. They were shown in Neptune Court, the section of the museum that houses the triumphant *Suhaili*.

Tacita Dean also had four words from the very end of Donald Crowhurst's logbook carved, in his handwriting, into a wooden guardrail at Neptune Court. The location provides the clearest view of the distance between human aspiration and fallibility: one can now stand at that rail, looking down at *Suhaili*—not a boat's length away—and read between one's hands:

"IT IS THE MERCY."

Acknowledgments

EACH BOOK IS ITS OWN PECULIAR VOYAGE. This one has been marked by the people who have helped me.

My closest shipmate has been my editor Dan Conaway. Dan sees forest and trees, and allows no arrangement of twigs to lie unexamined. His obtuse squiggles, withering use of the word "quaint," nearly illegible scrawlings all over the pages of every draft, and his ability to put his finger on what is wrong and show me what would be right, have raised the quality of the book well above what I might have been happy with. I'm profoundly grateful to him not only for all this, but also for his consistent grace in dealing with me when I have been less gracious with him.

Behind every great editor is an assistant, largely unsung, whose job is long, brutal, and unglamorous. This is the seven-tenths of the iceberg that is the foundation of seeing the finished manuscript into print. Dan's assistant Nikola Scott did this with unrelenting enthusiasm. She also made valuable points that materially improved the book.

Martha Cameron copy-edited the book. I'm thankful for her eye, ear, and erudition.

Andrew Franklin of Profile Books, London, provided me

with page by page comments that helped the book. Nicky White and Kate Griffin at Profile have felt like partners through two books. Nicky located, bought, and sent me research books that I could not have done without.

This business is not easy. My agent and friend Sloan Harris is courtly, honest, and a passionate advocate. He makes it seem possible and gives me courage. Thanks to Teri Steinberg.

Jonathan Raban urged me to dig deeper in certain areas, advice that had an incalculable effect of improving the whole book, and I'm very grateful to him.

Sam Manning's enthusiasm and ability to produce the maps I see in my head, but better, have been a boon to two of my books.

Tacita Dean shares my obsession with parts of this story. She has been generous with her work and thoughts and insights, and has allowed me to use her electrifying photograph of *Teignmouth Electron*. Counting her a friend has been one of the unanticipated joys of this book. Thanks also to Dale McFarland and all at the Frith Street Gallery, London.

Charlotte Brown at News International, London, helped me find my way through the *Sunday Times* photo archives. This was in a subterranean cavern in London's East End that had the homey feel of wartime Britain; the sort of place that needed a real, old-fashioned archivist with a sixth sense for the crucially misplaced, and that was Charlotte.

Thanks also to:

Derek Kelsall for his comments on multihulls and Nigel Tetley; Don Love at Production International helped me with a video, as did Rory Healy of the BBC;

Matt Murphy at *WoodenBoat* magazine gave me access to his magazine's incredible library, a gem from which lies in my book; Jon Wilson, Matt Murphy, and all at *WoodenBoat* have changed the world in an important way, and made life richer for me, and tens of thousands of others; Steve and Laurie White of Brooklin, Maine, made my stay there a happier one; Joel White was a deep influence, and will always be;

Cynthia Hartshorn on Cape Cod; Chris and Petey Noyes in Maine; Penny and Robert Germaux, Frank Field, Harriet Guggenheim in Spain; Irina Zamorina in New York; Greg and Sara Johnson, and Doug Grant and Kathryn Van Dyke in Mill Valley; Howard Sharp in wildest Canterbury; and always Annie Nichols;

Betsy Beers for humor and wisdom; Carole Fungaroli for putting me in the canon, and being my friend and most valuable resource at Georgetown University;

Marion and Jeric Strathallan for giving me a home in London, twice, greatly facilitating my research there; Mary Elliot for her room; my mother, Barbara Nichols, for that peculiarly right place in damp, rainy Spain where I've now written big chunks of three books; Liz and Tony Sharp for two productive stays in Mallorca; Joan deGarmo for a haven between incarnations; David Nichols for belief and encouragement; Matt and Sheila deGarmo for being uncomplaining and generous hosts during too much coming and going.

Matt has made so much possible for me over the years since I staggered ashore shipwrecked that he deserves more than just a mention at the back of the book.

Sara Nelson has been a good friend during the writing of this book. Thanks for the cards.

Sources

The London *Sunday Times*
The Daily Mirror
The Sunday Mirror
The Kent Messenger
J. R. L. Anderson, *The Ulysses Factor*
Chay and Maureen Blyth, *Innocent Aboard*
Charles A. Borden, *Sea Quest*
Francis Chichester, Gypsy Moth IV *Circles the World*
Adlard Coles, Peter Bruce, *Heavy Weather Sailing, 4th Edition*
Tacita Dean, *Teignmouth Electron*
Richard Henderson, *Singlehanded Sailing, 2nd Edition*
Eric Hiscock, *Voyaging Under Sail,* etc.
Hydrographic Office of the British Navy, *Ocean Passages for the World*
Bill King, *Capsize*
Robin Knox-Johnston, *A World of My Own*
Bernard Moitessier, *The Long Way; Tamata and the Alliance; Cape Horn, the Logical Route*
Jonathan Raban, *The Oxford Book of the Sea*
John Ridgway, *The Road to Ardmore*

Nigel Tetley, *Trimaran Solo*
Nicholas Tomalin and Ron Hall, *The Strange Last Voyage of Donald Crowhurst*

Permissions

Grateful acknowledgement is made for permission to reprint excerpts from the following copyrighted works:

From *A World of My Own* by Robin Knox-Johnston, copyright © 1969 by Robin Knox-Johnston. Used by permission of W. W. Norton & Company, Inc., and reproduced with permission of Curtis Brown Ltd, London, on behalf of Robin Knox-Johnston.

Bernard Moitessier, translated by William Rodarmor, *Tamata and the Alliance,* Dobbs Ferry, NY: Sheridan House Inc., 1995.

Bernard Moitessier, translated by William Rodarmor, *The Long Way,* Dobb's Ferry, NY: Sheridan House Inc., 1995.

Nicholas Tomalin and Ron Hall, *The Strange Last Voyage of Donald Crowhurst,* copyright © 1995 by the authors. Reproduced by permission of Hodder and Stoughton Limited and The McGraw Hill Companies.

All photographs copyrighted by the News International and the London *Sunday Times*, except for the photograph *Teignmouth Electron, Cayman Brac, 1999,* copyright © 1999 by

Tacita Dean, for which the author thanks Tacita Dean, the Frith Street Gallery, London, and the Marian Goodman Gallery, New York; and photograph of *Suhaili*, taken by Peter Nichols.

All maps by Samuel F. Manning, copyright © 2001.